Performance Metrics for Haptic Interfaces

Springer Series on Touch and Haptic Systems

Series Editors

Manuel Ferre
Marc O. Ernst
Alan Wing

For further volumes:
www.springer.com/series/8786

Evren Samur

Performance Metrics for Haptic Interfaces

 Springer

Evren Samur
Robotic Systems Laboratory
École Polytechnique Fédérale
   de Lausanne (EPFL)
Lausanne, Switzerland

ISSN 2192-2977   ISSN 2192-2985 (electronic)
Springer Series on Touch and Haptic Systems
ISBN 978-1-4471-5803-5   ISBN 978-1-4471-4225-6 (eBook)
DOI 10.1007/978-1-4471-4225-6
Springer London Heidelberg New York Dordrecht

Printed on acid-free paper

Springer is part of Springer Science+Business Media (www.springer.com)

To the women of my life:
Annem, Ablam, Elif and Ceren

Series Editors' Foreword

This is the fifth volume of the "Springer Series on Touch and Haptic Systems", which is published in collaboration between **Springer** and the **EuroHaptics Society**. *Performance Metrics for Haptic Interfaces* is focused on evaluating the performance of haptic devices by physical and psychophysical metrics. This work represents a significant step forward in haptic interface standardization, which helps the broader dissemination of human interaction devices that include haptic feedback.

The double approach presented in this volume is consistent with the nature of haptic devices: on the one hand, these devices are defined in terms of engineering features such as controllers, actuators, sensors, etc. On the other hand, haptic devices are designed to directly interact with users. Therefore, the human interaction capability will determine the suitable performance measure for such devices.

The basis for this volume is the PhD thesis of Evren Samur who was the winner of the 2011 EuroHaptics Society PhD award. It was selected from a pool of many other excellent works on haptics research. This monograph is an excellent example of the state of the art in the use of engineering and psychophysics for the development of haptics, and as such is an important starting point for future advances in this field.

Manuel Ferre
Marc Ernst
Alan Wing

Preface

The purpose of evaluation procedures for haptic interfaces is to achieve both qualitative and quantitative statements on haptic rendering realism and performance. Since haptic technology is being increasingly used in computer games, surgical simulators, mobile phones etc., there is a need for defining standards for haptic applications. This book aims at meeting this need by establishing standard practices for the evaluation of haptic interfaces and by identifying significant benchmark metrics.

Towards this end, a combined physical and psychophysical experimental methodology is given in this book. First, the existing physical performance measures and device characterization techniques were investigated and described in an illustrative way. The physical characterization methods were demonstrated on a two degrees-of-freedom haptic interface. Second, a wide range of human psychophysical experiments were reviewed and the appropriate ones were applied to haptic interactions. The psychophysical experiments were unified as a systematic and complete evaluation method for haptic interfaces. Seven psychophysical tests were derived and implemented for three commercial force-feedback devices. Experimental user studies were carried out and applicability of the tests to a tactile feedback device was investigated. Finally, synthesis of both evaluation methods is also discussed.

The generic methodology provided in this book enables readers to evaluate the suitability of a haptic interface for a specific purpose, to characterize and compare devices quantitatively and to identify possible improvement strategies in the design of the system.

Chicago, USA   Evren Samur

Acknowledgements

I would like to thank my PhD advisor, Hannes Bleuler, for giving me the opportunity to work at EPFL. I feel privileged to have been working with you and your team, full of brilliant people all the time. I always admire your enthusiasm and deep knowledge on the world culture and languages. Many thanks to Roger Gassert for not only giving me an extensive feedback but also encouraging me during the tough times of my research. Your stimulating support helped me develop confidence in my research. I would like to express my sincere appreciation to Paolo Fiorini, Matthias Harders and Max Hongler, for their valuable comments, suggestions and constructive criticisms. I would also like to thank Ed Colgate for accepting me as a visiting scholar in his lab at the Northwestern University, Chicago where I learned so much in a short period. Your brilliant and always challenging ideas advanced my critical thinking.

I also owe gratitude to Thomas Moix, Ulrich Spaelter, Dominique Chapuis, Lindo Duratti and Wang Fei. It was an unforgettable experience to work with you all. I am truly grateful to Lionel Flaction, Pascal Maillard, Andrew Watson, Andrew Whyte, Anders Larsson and our oversea colleagues Josh Passenger and David Hellier. Many thanks also go to all my colleagues and close friends at the lab, especially Solaiman Shokur, Aleksey Gribovskiy, Ricardo Pérez Suaréz, Masayuki Hara, Ricardo Beira, Ali Şengül, Giulio Rognini, Jeremy Olivier and Laura Santos Carreras.

Together with my long-lasting (academic) friends, Ufuk Olgaç, Erk Subaşı and FC Meral, we hit the road from Ankara, passed by Istanbul and arrived in Switzerland or in Chicago. I hope we continue to walk together. My dear friends Luca Pozzoli, Simos Koumoutsaris, Mathieu Aurousseau and Danielle Ramseier; you made me feel like at home in Switzerland. And now in Chicago, I am lucky to have beloved Rogers Park residents Matt Runfola and Sandra Stone in my life. I have to express my enormous gratitude to all my friends in Turkey for the joyful times we are spending—the list is too long to mention all the names. Elif the sister and İlham the brother; you always cared about me and showed your love. *Anne* and *Abla*; I dedicate this book to you because of your deep love for me. You raised me, continuously support me and tirelessly waited for me to come back home. My final thanks go to Ceren with whom I share every single moment of my life with love.

Contents

List of Acronyms, Symbols, Greek Letters

Acronyms

Abs.	Absolute
Amp	Amplifier
ANOVA	Analysis of Variance
Co-Me	Computer Aided and Image Guided Medical Interventions
CPT	Counts Per Turn
CSIRO	Australia's Commonwealth Scientific and Industrial Research Organisation
D/A	Digital to Analog
DC	Direct Current
DOF	Degree(s) of Freedom
DL	Difference Limen (Difference or Differential Threshold)
EPFL	Ecole Polytechnique Fédérale de Lausanne
EPs	Exploratory Procedures
Eq.	Equation
Enc	Encoder
Fig.	Figure
fMRI	functional Magnetic Resonance Imaging
FFT	Fast Fourier Transform
GCI	Global Conditioning Index
GIE	Generalized Inertia Ellipsoid
JND	Just-Noticeable-Difference
HD	High Definition
ID	Index of Difficulty
IP	Index of Performance
IR	Infrared
IT	Information Transfer
I/O	Input Output
ISO	International Organization for Standardization
LDV	Laser Doppler Vibrometer
LFFs	Lateral Force Fields

LSRO	Laboratoire de Systèmes Robotiques
MT	Movement Time
n/a	not applicable/not available
NCCR	National Centre of Competence in Research
PC	Personal Computer
RL	Reiz Limen (Absolute Threshold)
RMS	Root Mean Square
SDR	Structural Deformation Ratio
TPaD	Tactile Pattern Display
USB	Universal Serial Bus
var	Variance
VE	Virtual Environment

Symbols

a	Intercept
A	Distance of Movement
b	Reciprocal of IP
B	Damping
c	Constant
d_i	Damping
dW	Workspace Derivative
D	Diameter
$F(\omega)$	Force
F_a^n	Discrete Input Force
F_c	Control Effort
F_d	Desired Force
F_d^n	Discrete Desired Force
F_{ee}	Force at the End Effector
F_h	Voluntary Human Muscle Force
$\mathbf{G}$	Generalized Inertia Ellipsoid Matrix
H	Size of Hole
H_f	Transfer Function between Output and Input Force
H_v	Transfer Function between Output and Input Velocity
i	Input Current
I	Reference Stimulus
$\mathbf{J}$	Jacobian Matrix
k	Stimulus Categories
k_i	Stiffness
K	Virtual Stiffness/Spring
M	Mass
$\mathbf{M}$	Mass Matrix
M_i	Mass/Inertia
n	Discrete Parameter Indicator/Total Number of Trials
$n_{i,j}$	Number of Joint Event
p	Probability
P	Size of Peg

q	Pose
$n_{i,j}$	Encoder Count
$\dot{q}$	Joint Rates
$\dot{q}_d^n$	Discrete Joint Rates
R	Sampled Output Sine
R_j	Number of Responses
s	Laplace Transform Variable
S	Shape Index
S_{cor}	Step Size for Correct Answer
S_i	Number of Stimuli
S_{incor}	Step Size for Incorrect Answer
S_{unsure}	Step Size for Unsure Answer
T_r	Rise Time
$v(\omega)$	Velocity
v_{ee}	Velocity of the End Effector
v_{ee}^n	Discrete Velocity of the End Effector
v_e^n	Discrete Velocity of Virtual Environment
V	Voltage
V_{in}	Voltage to Amplifier
W	Width of Target/Precision
y	True Sine
Y_d	Admittance of Device
Y_f	Modified Admittance of Device
$Z(\omega)$	Mechanical Impedance
Z_c^n	Impedance of Virtual Coupling
Z_d	Impedance of Device
Z_d^n	Discrete Model of a Device Impedance
Z_e^n	Impedance of Virtual Environment
Z_h	Impedance of Human Hand and Arm
Z_i	An Impedance
Z_E^n	Total Impedance of Virtual Environment and Virtual Coupling
Z_t	Total Impedance of Human Hand and Device

Greek Letters

ΔI	Differential Threshold
$\Delta I/I$	Weber Fraction
ΔV	Step Voltage
η	Global Conditioning Index
κ_2	Condition Number
μ	Manipulability Index/Coefficient of Friction
μ_i	Friction
ρ	Encoder Pulse
σ_m	Smallest Singular Value of Jacobian
σ_M	Largest Singular Value of Jacobian
τ_a^n	Discrete Input Joint Torque

τ_i^n	Discrete Input to D/A
τ_m	Joint Torque
ω_b	(Operating) Bandwidth

Part I
Basic Considerations

Chapter 1
Introduction

Abstract There is a growing need for defining evaluation standards for haptic applications. The purpose of evaluation procedures for haptic interfaces is to achieve both qualitative and quantitative statements on haptic rendering realism and performance. However, it is challenging to define a performance indicator since technical device evaluation is not straightforward to interpret the results in terms of perceived rendering quality. While a haptic device may come with specifications for its mechanical and electrical properties, no clear relationship between these properties and application-specific performance is available from neither commercial nor academic literature. In this chapter, similar problems of haptic interface performance evaluation are investigated in depth and the objectives of this book are provided.

1.1 Motivation

A *haptic interface* is an actuated, computer controlled and instrumented device that allows a human user to touch and manipulate objects either within a virtual environment (VE) [3] or in a real world through a slave of a teleoperated systems such as for surgical robotics. The haptic interface ensures bilateral interactions between the user and the VE in a haptic rendering process as shown in Fig. 1.1. This dual way property, in other words being not only an input interface but also a feedback source for the user, gives a unique characteristic to the haptic device. The quality of a haptic device hardware and controller design directly affects realism, presence and immersion in a VE.

Haptic interface evaluation is necessary in order to make both quantitative and qualitative statements on rendering realism, performance and enhancement. *Performance of a haptic interface*, in a broad sense, can be defined as its ability to render a wide range of haptic stimuli. However, it is challenging to define an indicator of performance since technical device evaluation (i.e. bode-diagram, bandwidth) is not straightforward to interpret the results in terms of *perceived rendering quality* (i.e. "feel").

For any user interface such as computer screens, audio speakers or computer mouses, the user chooses the best performing one according to its specifications defined based on standards. These standards for audio and video interfaces have been defined in detail making comparison easy. However, this is not the case for haptic

E. Samur, *Performance Metrics for Haptic Interfaces*,
Springer Series on Touch and Haptic Systems,
DOI 10.1007/978-1-4471-4225-6_1, © Springer-Verlag London 2012

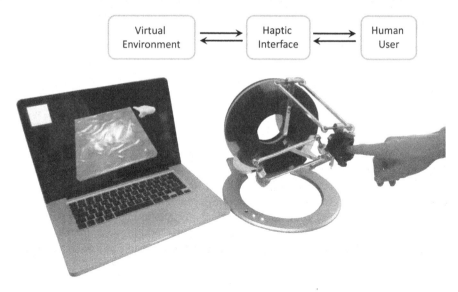

Fig. 1.1 Haptic rendering process. Bilateral interactions between a user and a virtual environment are realized through the haptic interface

interfaces. There is a growing need for defining evaluation standards for haptic applications such as those focused on surgery simulations, computer games, steer-by-wire systems of next-generation cars, etc. Which haptic device is better than the other? How to compare them based on some golden standard?

Although a set of performance metrics for haptic interfaces has been defined by Hayward and Astley [4], only few of them are generally provided by designers or manufacturers. That is because some of these metrics (e.g. frequency response and bandwidth) are difficult to quantify or measure due to dependency on boundary conditions and control. Besides, it is almost impossible to find detailed information on specifications, testing conditions and methods. Therefore, it is misleading to compare these specifications since there is no consensus on measurement methods which vary considerably between studies. Uncertainty and inconsistency in the measurement methods arise from the fact that a haptic interface is meant to be used by a human user. Its performance is highly affected by the user behavior and constraints. Therefore, any measurement performed without human interaction does not really represent the actual device use. On the other hand, human-in-the-loop physical experiments are subjected to a high level of disturbance that makes them not repeatable enough. Due to these unstandardized boundary conditions and measurement methodologies, the specifications given (if ever!) by the manufacturers are not comparable.

Application based or task specific evaluation methods have been proposed to test haptic interfaces during their proper use. This approach requires a human user to perform the task, thus it involves human dynamic and intention uncertainties. Although there are a variety of task specific evaluation approaches for haptic interfaces, it has not yet been possible to define a norm for meaningful device comparison and assessment. An evaluation procedure for haptic interfaces should link device performance measures to the limits of human perception in order to obtain device-specific limits. Besides, it needs to be simple enough to be easily implemented while taking all important attributes of haptic interaction into account.

Considering the increased number of psychophysical tests, ergonomics studies etc. performed with commercial haptic interfaces to reveal human sensory and motor control abilities, any experimenter performing this kind of studies must be well aware of the device limitations and capabilities. The resolution limits, fidelity and other metrics of a haptic device significantly influence the results of human threshold experiments. In technical terminology, *fidelity* is defined to be the level of exactness in the effect of a device, comprising also accuracy and precision. Weisenberger et al. [6] suggested that if two devices have different fidelities, then it is possible that differences in sensory thresholds might be attributable to the device rather than to the user's perceptual system. The fidelities and limits should be quantified in the frame of a standardized method. Otherwise, results of such a psychophysical study might be misleading.

1.2 Objectives and Approach

The ultimate goal of this book is to establish a norm for evaluation of haptic interfaces with significant benchmark metrics. This norm should be as generic as possible to enable meaningful and fair device comparison and assessment. It should also provide new quantitative performance metrics which are easier to be linked with the rendering quality perceived by the user or, in other words, the actual "feel" of the device.

In order to better understand the problems and challenges described in the previous section, existing physical performance measures and methodologies are categorized and described in an illustrative way. Thus, obtained tutorial-like guidelines for physical device evaluation describe testing conditions and methods with enough detail to perform identical tests with other devices. The physical characterization method is experimentally demonstrated on a two degrees-of-freedom haptic interface for colonoscopy simulation.

Evaluation procedures for haptic rendering must take into account all components of haptic interaction, i.e., the virtual reality environment, the haptic interface and the user behavior. Therefore, the human should be in the loop not only physically but also cognitively during the evaluation process. This necessity directs our attention also to usability evaluations. *Usability* is the ease of use and usefulness, including quantifiable characteristics, such as user task performances, subjective satisfaction, learnability and user comfort [2]. *Usability evaluation* is defined as the assessment of a specific application's user interface, an interaction technique, or an input/output device, for the purpose of determining its actual or probable usability. One of the main aspects of VE technology is being able to have multiple inputs and outputs, such as visual, audio and haptic modalities. Therefore, usability evaluation of such input/outputs is crucial in order to design effective VEs with augmented presence, immersion and system comfort [5]. Although usability studies have been extensively applied to traditional graphical user interfaces, there are only few evaluation processes for haptic interfaces. In order to quantify how well an existing interface supports haptic interaction, we need to perform usability evaluations with haptic interfaces.

As haptic interfaces are intended to be used by a human user, human psychophysical limits provide minimum requirements for the haptic interaction. Since these limits are well studied and documented in the literature, human perception can also be used to evaluate the haptic interfaces. Hence, our approach is to apply human performance estimations to haptic interactions in order to have a systematic and complete evaluation method for haptic interfaces. A wide range of different psychophysical experiments are reviewed and investigated. Based on these resources, we apply Bowman and Hodges' testbed evaluation approach [1] to haptic interactions and synthesize a set of evaluation testbeds. Careful repetition of described human factor studies leads to basic quantitative benchmark metrics for haptic interfaces. This approach allows us to evaluate force-feedback devices in a generic context to judge the quality of their design in a quantitative way. Validity of the tests for force-feedback interfaces are experimentally proven through several user studies with three commercially-available devices. In addition, feasibility of the tests for tactile interactions are also investigated.

1.3 Outline

State of the art and useful background information can be found in Chaps. 2 and 3. In Chap. 3, psychophysical studies on human haptic perception are reviewed. Basic properties of the hand and psychophysical methods are discussed to provide background information. In order to understand the human sensory performance, the haptic perception limits are summarized which are then used in the following chapters as a reference.

Chapter 4 presents the physical performance evaluation methodologies and metrics. First, a model of haptic interface performance is provided. Then, the methodologies are categorized based on this model and described in detail. Finally, these methods are applied to a 2-DOF haptic interface and the results of physical evaluation are provided in Chap. 5.

A psychophysical evaluation methodology is presented in Chap. 6. The testbed evaluation approach is applied to haptic interactions and a set of benchmark metrics are provided for haptic interfaces. We describe the methodology of seven testbeds and experimentally demonstrate usefulness of the testbeds on three commercial force-feedback devices. Some directions for new designs based on psychophysics are also indicated. The application of the testbeds to a tactile feedback device is investigated in Chap. 7.

Finally, a synthesis of physical and psychophysical evaluation methods and the contributions of this book are summarized in Chap. 8. Possible future extensions of our research are also discussed in the outlook section.

References

1. Bowman, D.A., Hodges, L.F.: Formalizing the design, evaluation, and application of interaction techniques for immersive virtual environments. J. Vis. Lang. Comput. **10**(1), 37–53 (1999)
2. Bowman, D.A., Gabbard, J.L., Hix, D.: A survey of usability evaluation in virtual environments: classification and comparison of methods. Presence: Teleoperators Virtual Environ. **11**(4), 404–424 (2002).
3. Gillespie, R.B.: Haptic interface to virtual environments. In: Kurfess, T. (ed.) Robotics and Automation Handbook. CRC Press, Boca Raton (2005)
4. Hayward, V., Astley, O.: Performance measures for haptic interfaces. In: Robotics Research: The 7th International Symposium, pp. 195–207 (1996)
5. Stanney, K.M., Mollaghasemi, M., Reeves, L., Breaux, R., Graeber, D.A.: Usability engineering of virtual environments (VEs): identifying multiple criteria that drive effective VE system design. Int. J. Hum.-Comput. Stud. **58**(4), 447–481 (2003)
6. Weisenberger, J., Kreier, M., Rinker, M.: Judging the orientation of sinusoidal and square-wave virtual gratings presented via 2-dof and 3-dof haptic interfaces. Haptics-e **1**(4) (2000)

Chapter 2
State of the Art

Abstract Haptic interfaces generate the sense of touch in the form of force or tactile feedback and allow us to touch and manipulate objects either within a virtual environment or in a real world through a slave of a teleoperated system, such as for surgical robotics. There has been considerable amount of research on the haptic technology, which brought it into computer games, surgical simulators, mobile phones etc. A closer investigation of these devices and studies on their performance evaluation shows that type of evaluations, aim of methods and performance metrics vary considerably depending on the device. We have, therefore, reviewed the evaluation methods in the literature that have been applied to haptic devices. In this chapter, first, commercially available haptic interfaces and their application areas are reviewed. Then, haptic interface evaluation studies in the literature are discussed and categorized into two groups: physical and psychophysical evaluation studies.

2.1 Haptic Interfaces

Haptic technology deals with the synthesis of touch and force (*haptics*, in general) to enable us to interact with virtual environments through haptic interfaces. In short, haptic interfaces generate the sense of virtual touch in the form of force feedback (for receptors in the muscles and joints) and tactile feedback (for sensors located in the skin). Since the early 1990s, there has been considerable amount of research on the haptic technology which brought it into computer games, surgical simulators, mobile phones etc. The following section reviews the state-of-the-art haptic interfaces.

2.1.1 Force Feedback Devices

2.1.1.1 General Purpose Interfaces

Perhaps the most widely used haptic interface is the PHANTOM® developed by Massie and Salisbury [52] and now commercialized by the Sensable Technologies,

E. Samur, *Performance Metrics for Haptic Interfaces*,
Springer Series on Touch and Haptic Systems,
DOI 10.1007/978-1-4471-4225-6_2, © Springer-Verlag London 2012

Fig. 2.1 PHANTOM Omni®
from Sensable Technologies,
Inc.®

Fig. 2.2 The omega.3 haptic
device from Force
Dimension. This 3-DOF
desktop interface with its
embedded real-time
controller offers a universal
interface for standard haptic
applications. Photo courtesy
of Force Dimension

Inc.® [74]. Different versions are available, ranging from a low cost desktop application (PHANTOM Omni®, 0.055 mm position resolution, peak force 3.3 N) to a
high-end research tool (PHANTOM Premium 1.5 High Force/6-DOF, 0.007 mm position resolution, 37.5 N peak force). As shown in Fig. 2.1, the stylus of PHANTOM
Omni enables position and orientation input in 6-DOF while force feedback is only
in 3-DOF. Recently, a new handle design for the 6-DOF family of haptic devices
permits attaching interchangeable new end effectors providing pinch functionality.

The omega.x, delta.x and sigma.x haptic devices from Force Dimension [25] are
of the high performance interfaces. For example, the omega.3 is a 3-DOF desktop
interface allowing 12.0 N maximum continuous force feedback with a position resolution of 0.01 mm. Its parallel kinematics (see Fig. 2.2) design enables the omega.3
base to accommodate various interchangeable end-effectors to upgrade to multi-
DOF versions. On the other hand, delta.6 is more suitable for various engineering
applications and experimentations with its higher workspace and force feedback
capability in translational and rotational DOFs (see Fig. 2.3). Due to it is parallel
delta structure, it can generate high continuous forces and torques up to 20 N and
0.150 Nm. Finally, the recently released sigma.7 introduces seven active DOFs including grasping force feedback up to ± 8 N. This high-end haptic device as shown

Fig. 2.3 The delta.6 haptic device from Force Dimension. With its large workspace and active wrist end effector, the delta.6 is suitable for virtual reality based research and engineering. Photo courtesy of Force Dimension

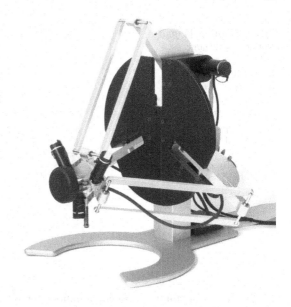

in Fig. 2.4, which has a maximum continuous force and torque of 20 N and 0.4 Nm respectively, is mainly used in the aerospace and medical fields demanding safety-critical applications. The Novint Falcon® (Novint Technologies, Inc.) is a low-cost version of the omega.3 targeting gaming industry, with a peak force around 10 N.

The HapticMaster [86] is the only admittance controlled haptic interface on the market (commercialized by MOOG, Inc. [57]). The admittance control enables it to achieve high force output (max continuous and peak force of 100 and 250 N) and render high impedance. Its large workspace and high impedance characteristics make this device an ideal candidate for the rehabilitation research.

Haption SA provides a wide range of haptic interfaces called Virtuose™ [34]. For example, the Virtuose 6D Desktop used for gaming applications has a maximum continuous force of 3 N, on the other hand, larger version of this device, MAT 6D,

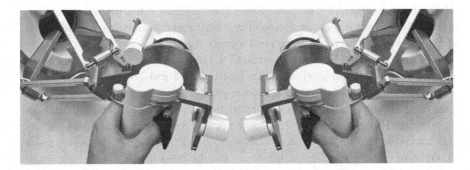

Fig. 2.4 The sigma.7 haptic device from Force Dimension. Being the most advanced haptic interface developed by Force Dimension, it introduces 7 active DOFs, including grasping capability. Photo courtesy of Force Dimension

Fig. 2.5 The 6-DOF
VIRTUOSE 6D35-45 from
Haption SA. Its large
workspace corresponding to
the movements of a human
arm and 6-DOF force
feedback, make it especially
suited for one to one virtual
object manipulation. Photo
courtesy of Haption SA

can generate a maximum of 30 N continuous force and used for teleoperation. Figure 2.5 shows a Virtuose 6D35-45 device which has a workspace corresponding to the movements of a human arm.

The Freedom 7S is a serial force feedback device designed by Hayward et al. [38] and later commercialized by MPB Technologies Inc. [59]. The device is especially designed for medical simulation. Although the maximum continuous force is not high (0.6 N), it has a position resolution of 0.002 mm which makes it suitable for precise applications.

For those looking for a high fidelity desktop device, Quanser Inc. provides two haptic interfaces [66]. First one is the 5 DOF Haptic Wand which is originally designed by the group of Prof. Tim Salcudean at the University of British Columbia, Canada [77]. The haptic interface allows for three translations and two rotations (roll and pitch) by using a dual-pantograph arrangement (see Fig. 2.6). Second haptic interface developed by Quanser Inc. is the 6 DOF High Definition Haptic Device (HD^2) shown in Fig. 2.7. Compared to the Haptic Wand, it has not only one additional DOF but also higher force capability (maximum continuous force and torque of 11 N and 0.950 Nm respectively) and a larger workspace.

Maglev 200™ from Butterfly Haptics, LLC [12] is the only commercially available magnetic levitation haptic interface (see Figs. 2.8 and 2.9). The haptic device employs the principles of Lorentz levitation which eliminates the drawbacks of systems using mechanical elements such as friction, backlash, link bending, and motor cogging. This gives Maglev 200™ superior performance characteristics such as zero backdrive friction, high force bandwidth (2 kHz) and high position resolution (0.002 mm) and high stiffness (50 N/mm). On the other hand, its relatively small workspace (24 mm diameter sphere) limits its application areas. The first generation magnetic levitation haptic device was developed by Prof. Ralph Hollis and his student Peter Berkelman at Carnegie Mellon [8].

Fig. 2.6 The 5 DOF Haptic Wand, developed by Quanser Inc. and Prof. Tim Salcudean of the University of British Columbia, Canada, is an open architecture solution designed to help research or teach haptics. Photo courtesy of Quanser Inc.

The specifications of the reviewed commercially available force feedback devices are summarized in Table 2.1. As shown in this table, not all the specifications are provided by the manufacturers. Although some specifications such as workspace and continuous force are common, important information on force resolution, transparency and frequency response characteristics is rarely provided.

2.1.1.2 Surgery Simulators

The positive impact of haptic feedback in virtual reality based surgery simulators has been recently proven by clinical trials [3, 5]. This is the reason why haptic interfaces

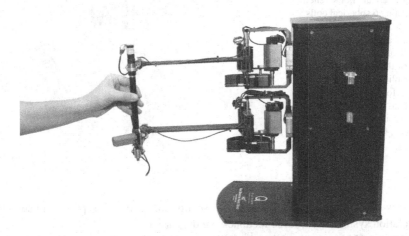

Fig. 2.7 The 6 DOF High Definition Haptic Device (HD^2), developed by Quanser Inc., is a high fidelity force-feedback platform for advanced research in haptics and robotics. Photo courtesy of Quanser Inc.

Fig. 2.8 Maglev 200™ Magnetic Levitation Haptic Interface is the first commercial haptic device to employ the principles of Lorentz levitation. The handle can be freely moved in 6 DOF with zero friction. Photo courtesy of Butterfly Haptics, LLC

Fig. 2.9 Maglev 200™ cut away picture: (**1**) Handle (or manipulandum), (**2**) Hemispherical "flotor" shell containing 6 spherical coils, (**3**) One of 6 permanent magnet assemblies, (**4**) One of 3 light emitting diodes, (**5**) One of three optical sensor assemblies, (**6**) Flexible wiring for power and signals, (**7**) Interface to controller. Photo courtesy of Butterfly Haptics, LLC

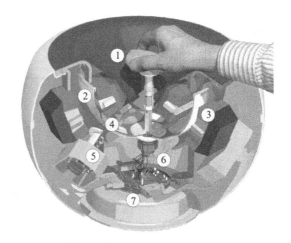

are very successful in this domain. In this section, the companies providing surgery simulation systems with force feedback are discussed.

Mentice [55] uses the Xitact™ IHP in their laparoscopy simulator (Mentice MIST™). The Xitact IHP is a 4-DOF force feedback device which was originally developed by Dr. Vollenweider at Ecole Polytechnique Fédérale de Lausanne [89].

Table 2.1 Specifications provided by the manufacturers of the commercially available force feedback devices

	Phantom Omni [74]	omega.3 [25]	Novint Falcon [62]	Haptic Master [86]	Virtuose [34]	Freedom 7S [59]	HD^2 [66]	Xitact IHP [55]	Maglev 200 [12]
Structure	Serial	Parallel	Parallel	Serial	Serial	Serial	Parallel	Hybrid	n/a
Workspace (10^{-3} m^3)	1.3	2.2	1.1	80.0	91.0	12.0	70.0	1.9	0.01
DOF	6 // 3	3	3	3	6	7	6 // 5	4	6
Position Res. (µm)	55.0	10.0	60.0	4.0	6.0	2.0	51.0	57.0	2.0
Continuous Force (N)	0.9	12.0	9.0	100.0	10.0	0.6	10.8	20.0	n/a
Peak Force (N)	3.3	n/a	n/a	250.0	35.0	2.5	19.7	30.0	40.0
Stiffness (N/mm)	2.0	14.5	n/a	50.0	n/a	2.0	3.0	n/a	50.0
Stiction (N)	0.26	n/a	n/a	n/a	n/a	0.04	0.35	n/a	0.0
Force Resolution (N)	n/a	n/a	n/a	0.01	n/a	n/a	n/a	n/a	0.02
Force Bandwidth (Hz)	n/a	n/a	n/a	n/a	n/a	n/a	n/a	n/a	2000

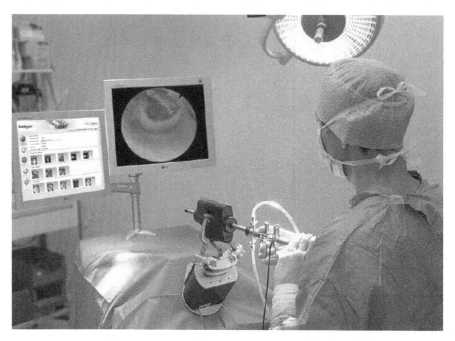

Fig. 2.10 VirtaMed HystSim system employing the Xitact™ IHP as the force feedback interface. Photo courtesy of VirtaMed AG

In addition, Mentice has an endovascular simulator called VIST™ that enables force feedback.

Virtamed [88] is a Swiss start-up company producing HystSim system originally developed by Harders et al. [36]. It enables training of diagnostic and therapeutic hysteroscopy using an original resectoscope and provides objective performance feedback. The prototype of the HystSim used to work with a haptic interface developed at Ecole Polytechnique Fédérale de Lausanne [76]. Currently, it uses the Xitact™ IHP as a force feedback device (see Fig. 2.10).

CAE Healthcare [13] has three simulators that provide force feedback: the LaparoscopyVR, the EndoscopyVR and CathLabVR. The LaparoscopyVR is designed for teaching minimally invasive laparoscopic surgery and force feedback is provided by a 3-DOF device. The EndoscopyVR (formerly AccuTouch System® of Immersion Corp.) is a simulator for teaching and assessing motor skills for gastrointestinal and bronchial assessment. It provides 1-DOF force feedback during insertion and removal of the endoscope. Finally, CAE Healthcare's CathLabVR simulates vascular procedures with force feedback.

Surgical Science [78] has a laparoscopy simulator (LapSim) which is also compatible with the Xitact™ IHP. In addition, together with Ecole Polytechnique Fédérale de Lausanne (EPFL) and Commonwealth Scientific and Industrial Research Organisation (CSIRO), they are developing an endoscopy simulator which includes a 2-DOF force feedback device [70] (see Fig. 2.11).

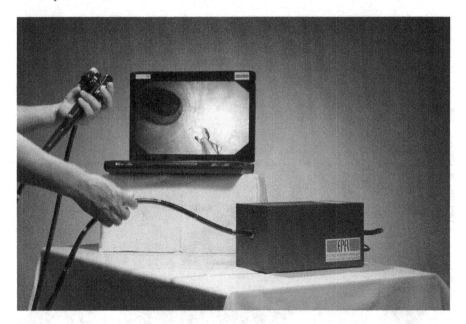

Fig. 2.11 The haptic interface developed at EPFL [70] integrated with the software simulation for colonoscopy (MILX™ GastroSim) developed at CSIRO [19, 39]. The simulator is currently being commercialized by Surgical Science Sweden AB

Simbionix USA Corporation [75] recently introduced the GI-BRONCH Mentor™ a combined platform for GI endoscopy and flexible bronchoscopy. This simulator provides higher force feedback by a pneumatic balloon breaking system, yet the translational and rotational force feedback are not decoupled. Simbionix also offers a laparoscopy simulator (LAP Mentor™) with force feedback.

Mimic Technologies [56] has developed a training simulator (the dV-Trainer™) designed for training of surgeons learning to use da Vinci® Surgical Robotic System from Intuitive Surgical®, Inc. The haptic interface is a novel cable driven system [9].

2.1.1.3 Surgical Robotics

Robotic surgery has been a domain of intense research activity in recent years. Despite the certain benefits such as providing high-definition visualization system and enhanced dexterity, the use of a teleoperated robotic system removes the direct contact of hands with tissues and thus, diminishes the sense of touch. All information about the patient is given to surgeons only through the visual sense. This imposes surgeons to exclusively rely on visual cues, compromising patient safety and telepresence. From the surgeons' perspective the force feedback plays a crucial role for patient safety and intuitiveness [71]. However, up to now, the potential of haptic feedback in robotic surgery has not yet been fully exploited and thus, this application still represents a fascinating research field.

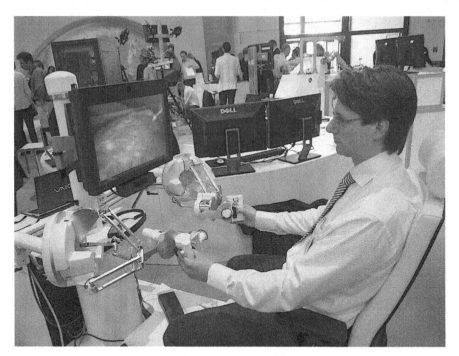

Fig. 2.12 DLR's MiroSurge haptic console consists of two sigma.7 haptic device from Force Dimension. Photo courtesy of Force Dimension

Intuitive Surgical's da Vinci® Surgical System [43] is the leading surgical robot which is used in several operations such as urology, gynecology and general surgery. This system provides 3D steroscopic vision and high dexterity to control the surgical instruments at the tip of the robot. However, force feedback resulting from interaction between the instruments and tissues is neglected and surgeons using this system rely on only visual cues [29]. Since it has been shown that force feedback enhances performance in robotic surgery [45, 72, 73, 90], there have been several efforts to restore the sense of touch when using the da Vinci system. For example, the VerroTouch [54] measures the impact caused by tool contacts inside patients and reproduces them at the level of the master handle. This feedback allows the surgeon to feel important tactile events such as rough surfaces as well as the beginning and the end of contact during manipulations. King et al. [45] developed a tactile feedback system to translate force distribution on the da Vinci surgical instruments to the fingers. In parallel to direct feedback, sensory substitution with imaging techniques is also proposed for restoring haptic feedback [47]. Nevertheless, this extra information should always be introduced carefully to avoid mental (or visual) overload. Despite these efforts to overcome the lack of haptic feedback in the da Vinci system, the proposed methods are still far from being perfect and not available in the commercial version.

Fig. 2.13 Aphee-4x
pin-based tactile interface
from Aesthesis. Photo
courtesy of Aesthesis

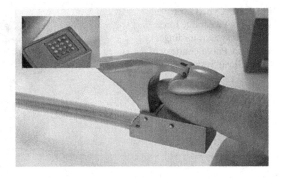

The only commercially available tele-operated surgical robotic system with haptic feedback is the Sensei™ X Robotic Catheter System from Hansen Medical, Inc. [33]. This robotic catheter system uses the 3-DOF omega.medical haptic device from Force Dimension to control the tip of the catheter. Force feedback information based on preoperative data is provided to the surgeon in real time, while maintaining patient safety.

Force Dimension has recently developed the sigma.7 haptic device [25] which is dedicated for medical applications. MiroSurge surgical robot from German Aerospace Center (DLR) [30, 82] features two sigma.7 haptic devices which have force feedback in 7 degrees of freedom including grasping (see Fig. 2.12). However, the MiroSurge is not yet commercially available.

2.1.2 Tactile Interfaces

Contrary to vast number of force feedback devices on the market, there are not many commercially available tactile interfaces. Until couple of years ago, pin-based tactile interfaces were quite common. One example to this kind of tactile devices is the Aphee-4x from Aesthesis [2]. This interface consists of an array of 16 fingertip pins arranged in an area of 7 mm^2 and can reproduce surface profiles of virtual objects on the fingertip as shown in Fig. 2.13.

The tactile technology has recently found his common application in mobile phones and gaming interfaces as simple vibrating buzzes. Nowadays, almost all mobile phones have a vibrating mode. Nintendo Wii [61] and Logitech Driving Force™ GT [51] are two examples of tactile interfaces used in computer games for better realism and immersion.

Now tactile technology in touch screens and mobile phones is going beyond the primitive haptics and presenting the boundaries or surface properties of an object on screen as you move your finger over it. TouchSense® tactile technology from Immersion Corp. [42] is claimed to provide "HD haptics" using piezo actuators. This technology is already integrated in Immersion's touch screens and some mobile phones such as Synaptics Fuse [79]. It is also used in cars to facilitate drivers to select an icon on the control menu.

Fig. 2.14 CyberGrasp
system is a wearable force
feedback system for fingers
and hand. Photo courtesy of
CyberGlove Systems LLC

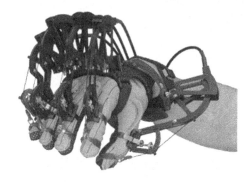

2.1.3 Other Applications

Apart from the haptic devices mentioned above, there are also other application areas worth mentioning. Introduction of robotic systems into the area of stroke rehabilitation has improved the therapy outcome. For example, Hocoma AG has several rehabilitation robotic systems which utilize force feedback for locomotion therapy (Lokomat®) and functional therapy of the upper extremities (Armeo®) [40].

In addition to the grounded desktop force feedback devices mentioned earlier, force feedback gloves are also available for gaming and rehabilitation purposes. CyberTouch, CyberForce and CyberGrasp are three different wearable systems from CyberGlove Systems LLC with tactile or force rendering capability for each finger and hand [18]. The CyberGrasp device shown in Fig. 2.14 is a lightweight, force-reflecting exoskeleton that fits over a CyberGlove data glove and adds force feedback to each finger. Grasp forces (up to 12 N per finger) are produced by a network of tendons routed to the fingertips via the exoskeleton.

2.2 Evaluation Studies

A closer investigation of studies on the evaluation of haptic rendering shows that type of evaluations, aim of methods and performance metrics vary considerably in these studies. We have therefore categorized the evaluation methods in the literature that have been applied to haptic interactions including VE, control, device as well as the human operator (see Fig. 2.15). Some of these methods employ only algorithm validation and comparison based on rendering realism [50, 67], whereas some others studied control design and evaluation for haptic interfaces [10, 17, 31, 48].

2.2.1 Physical Evaluation Studies

The discussion about experimental performance evaluation for haptic interfaces goes back to 80s when force-reflecting hand controllers (today's haptic interfaces)

Fig. 2.15 Classification of haptic rendering evaluation techniques and their corresponding basic performance criteria

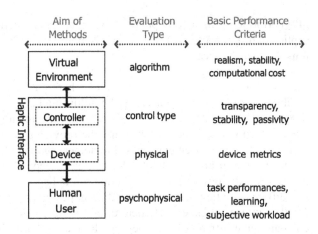

were used in teleoperation. The design requirements for teleoperation were described by Brooks [11] and used by many researchers. Later, McAffee and Fiorini [53] identified the key performance characteristics of the hand controllers and quantitatively compared existing devices. Hollerbach et al. [41] made a comparative analysis of actuator technologies for robotics. One of the detailed studies to measure force output performance of a robot was carried out by Eppinger [22]. He modeled the robot dynamic performance and conducted experiments to extract the effect of different components of the robotic system. Hayward and Astley [37] theoretically defined performance measures directed towards isotonic (i.e. impedance type) devices. More or less at the same time, these measures were formalized for coupled micro-macro actuators by Morrell and Salisbury [58]. In addition, practical ways to measure them were experimentally demonstrated on a haptic interface by Ellis et al. [21]. Several projects [1, 7, 23, 27, 65, 70, 87, 95, 96] evaluated particular haptic devices based on these technical performance metrics (these metrics are studied in detail in Chap. 4). An experimental identification method was described by Frisoli and Bergamasco [26]. Similarly, the dynamics of PHANTOM Premium 1.5A (Sensable Technologies Inc.) were experimentally identified by [14, 80]. Ueberle [84] conducted hardware experiments for the comparative performance evaluation of haptic control schemes using the VISHARD interface [83]. Weir et al. [93] described methods to measure impedance distribution of a haptic device over frequency based on the *Z-width* concept [17]. A method was proposed by Chapuis to calculate the output impedance of a device using the electrical analogy [15].

2.2.2 Psychophysical Evaluation Studies

There are many human factor studies to asses the benefits of haptic feedback on sensory-motor control tasks. Peg-in-hole [32, 85], tapping [16, 92], targeting [63], haptic training [4], joint tasks in a shared VE [6] and object recognition [64, 81, 94] tests are the most frequently performed experiments in these studies. Lawrence et al.

[49] performed some psychophysical experiments to ascertain whether human perception of differences in hardness depends more on high frequency or low frequency impedance differences.

In spite of the large number of psychophysical studies, only few of the tests have been used to measure the performance of a haptic interface rather than the haptic feedback itself. Wall and Harwin [92] employed a tapping test in conjunction with Fitts' law [24] in order to establish a measure of human performance in a simple target selection task. They showed that the providing force feedback significantly reduced subjects' movement times. In another study [91], they measured the performance of their high bandwidth device in a perceptual context of roughness [94] in order to fully evaluate its contribution to the haptic system. They demonstrated that different haptic interfaces have different performance characteristics in rendering the surface roughness. Harders et al. [35] performed 3D peg-in-hole tests to compare three different haptic devices. Rendering hard virtual walls has been the most mentioned benchmark topic in evaluating the performance of haptic interfaces. Lawrence et al. [49] introduced rate-hardness as a quality metric which is more relevant than mechanical stiffness in perception of hardness. Guerraz et al. [28] suggested to use physical data from a haptic device to evaluate haptic user interfaces. Kappers et al. [44] performed haptic identification experiments using quadric surfaces and showed that both shape index, a quantity describing the shape, and curvedness had significant effect on haptic shape identification. Based on this research, Kirkpatrick and Douglas [46] used shape recognition as an evaluation method for a complete haptic system. Their protocol can be used as a benchmark task to evaluate new haptic interface designs but it does not comprise all haptic interactions. Moreover, Tan [81] applied the absolute identification paradigm to sphere size identification for human performance estimations. Results were expressed in bits of information transfer and showed that humans could correctly identify at most 3 to 4 sphere sizes (corresponding to 2 bits) ranging from 10 to 80 mm in radius using the PHANTOM. This conclusion is also consistent with the results of manual length identification with physical objects [20], thus 2 bits of information transfer (IT) can be used as the threshold of identification performance of human for device evaluation. Murray et al. [60] used this information transfer concept to evaluate their wearable vibrotactile glove. Salisbury et al. [68, 69] used detection psychophysical experiments to measure device performance using vibrotactile stimuli. Their results indicated that none of the haptic devices tested were able to render perceptually distortion-free vibrations at detection threshold levels.

References

1. Adams, R.J., Hannaford, B.: Stable haptic interaction with virtual environments. IEEE Trans. Robot. Autom. **15**(3), 465–474 (1999)
2. Aesthesis: Aphee-4x. http://www.aesthesis.net/aphee-4x.html (2010)
3. Ahlberg, G., Enochsson, L., Gallagher, A.G., Hedman, L., Hogman, C., McClusky, D.A. III, Ramel, S., Smith, C.D., Arvidsson, D.: Proficiency-based virtual reality training significantly reduces the error rate for residents during their first 10 laparoscopic cholecystectomies. Am. J. Surg. **193**(6), 797–804 (2007)

4. Avizzano, C.A., Solis, J., Frisoli, A., Bergamasco, M.: Motor learning skill experiments using haptic interface capabilities. In: Proc. of 11th IEEE International Workshop on Robot and Human Interactive Communication, pp. 198–203 (2002)

5. Bajka, M., Tuchschmid, S., Streich, M., Fink, D., Szekely, G., Harders, M.: Evaluation of a new virtual-reality training simulator for hysteroscopy. Surg. Endosc. **23**(9), 2026–2033 (2009)

6. Basdogan, C., Ho, C., Srinivasan, M.A., Slater, M.: An experimental study on the role of touch in shared virtual environments. ACM Trans. Comput.-Hum. Interact. **7**(4), 443–460 (2000)

7. Bergamasco, M., Frisoli, A., Avizzano, C.: Exoskeletons as man-machine interface systems for teleoperation and interaction in virtual environments. In: Ferre, M., Buss, M., Aracil, R., Melchiorri, C., Balaguer, C. (eds.) Advances in Telerobotics. Springer Tracts in Advanced Robotics, vol. 31, pp. 61–76. Springer, Berlin (2007)

8. Berkelman, P.J., Hollis, R.L.: Lorentz magnetic levitation for haptic interaction: device design, performance, and integration with physical simulations. Int. J. Robot. Res. **19**(7), 644–667 (2000)

9. Berkley, J., Vollenweider, M., Kim, S.: Haptic systems employing force feedback. Patent Pub. No.:WO/2008/070584 (2008)

10. Botturi, D., Castellani, A., Moschini, D., Fiorini, P.: Performance evaluation of task control in teleoperation. In: Proc. of IEEE International Conference on Robotics and Automation (ICRA), vol. 4, pp. 3690–3695 (2004). doi:10.1109/ROBOT.2004.1308833

11. Brooks, T.L.: Telerobotic response requirements. In: IEEE International Conference on Systems, Man and Cybernetics, pp. 113–120 (1990)

12. Butterfly Haptics, LLC: Maglev 200™. http://butterflyhaptics.com/products/ (2012)

13. CAE Healthcare: EndoscopyVR. http://www.cae.com (2012)

14. Cavusoglu, M.C., Feygin, D., Tendick, F.: A critical study of the mechanical and electrical properties of the phantom haptic interface and improvements for high-performance control. Presence: Teleoperators Virtual Environ. **11**(6), 555–568 (2002). doi:10.1162/105474602321050695

15. Chapuis, D.: Application of ultrasonic motors to MR-compatible haptic interfaces. PhD thesis, EPFL, No. 4317 (2009)

16. Chun, K., Verplank, B., Barbagli, F., Salisbury, K.: Evaluating haptics and 3d stereo displays using Fitts' law. In: Proc. of the 3rd IEEE Workshop on HAVE, pp. 53–58 (2004)

17. Colgate, J.E., Brown, J.M.: Factors affecting the Z-width of a haptic display. In: IEEE Int. Conf. Robotics and Automation, pp. 3205–3210 (1994)

18. CyberGlove Systems LLC: CyberTouch. http://www.cyberglovesystems.com/ (2012)

19. de Visser, H., Passenger, J., Conlan, D., Russ, C., Hellier, D., Cheng, M., Acosta, O., Ourselin, S., Salvado, O.: Developing a next generation colonoscopy simulator. Int. J. Image Graph. **10**(2), 203–217 (2010)

20. Durlach, N.I., Delhorne, L.A., Wong, A., Ko, W.Y., Rabinowitz, W.M., Hollerbach, J.: Manual discrimination and identification of length by the finger-span method. Percept. Psychophys. **46**(1), 29–38 (1989)

21. Ellis, R., Ismaeil, O., Lipsett, M.: Design and evaluation of a high-performance haptic interface. Robotica **14**, 321–327 (1996)

22. Eppinger, S.D.: Modeling robot dynamic performance for endpoint force control. PhD thesis, MIT (1988)

23. Faulring, E.L., Colgate, J.E., Peshkin, M.A.: The cobotic hand controller: design, control and performance of a novel haptic display. Int. J. Robot. Res. **25**, 1099–1119 (2006)

24. Fitts, P.M.: The information capacity of the human motor system in controlling the amplitude of movement. J. Exp. Psychol. **47**, 381–391 (1954)

25. Force Dimension: Omega. http://www.forcedimension.com/ (2012)

26. Frisoli, A., Bergamasco, M.: Experimental identification and evaluation of performance of a 2 dof haptic display. In: Proc. of IEEE International Conference on Robotics and Automation, vol. 3, pp. 3260–3265 (2003)

27. Gassert, R., Moser, R., Burdet, E., Bleuler, H.: MRI/fMRI-compatible robotic system with force feedback for interaction with human motion. IEEE/ASME Trans. Mechatron. **11**(2), 216–224 (2006)

28. Guerraz, A., Loscos, C., Widenfeld, H.R.: How to use physical parameters coming from the haptic device itself to enhance the evaluation of haptic benefits in user interface? In: Proc. of Eurohaptics'03 (2003)

29. Hagen, M., Meehan, J., Inan, I., Morel, P.: Visual clues act as a substitute for haptic feedback in robotic surgery. Surg. Endosc. **22**, 1505–1508 (2008). doi:10.1007/s00464-007-9683-0

30. Hagn, U., Konietschke, R., Tobergte, A., Nickl, M., Joerg, S., Kuebler, B., Passig, G., Groeger, M., Froehlich, F., Seibold, U., Le-Tien, L., Albu-Schaeffer, A., Nothhelfer, A., Hacker, F., Grebenstein, M., Hirzinger, G.: DLR MiroSurge: a versatile system for research in endoscopic telesurgery. Int. J. Comput. Assisted Radiol. Surg. **5**(2), 183–193 (2010)

31. Hannaford, B., Ryu, J.: Time domain passivity control of haptic interfaces. IEEE Trans. Robot. Autom. **18**(1), 1–10 (2002)

32. Hannaford, B., Wood, L., McAffee, D., Zak, H.: Performance evaluation of a six axis generalized force reflecting teleoperator. IEEE Trans. Syst. Man Cybern. **21**, 620–633 (1991)

33. Hansen Medical, Inc.: Sensei™ X Robotic Catheter System. http://www.hansenmedical.com/ (2012)

34. HAPTION: Virtuose 6D35-45. http://www.haption.com/ (2012)

35. Harders, M., Barlit, A., Akahane, K., Sato, M., Szkely, G.: Comparing 6dof haptic interfaces for application in 3d assembly tasks. In: Proc. of Eurohaptics'06 (2006)

36. Harders, M., Bachofen, D., Bajka, M., Grassi, M., Heidelberger, B., Sierra, R., Spaelter, U., Steinemann, D., Teschner, M., Tuchschmid, S., Zatonyi, J., Szekely, G.: Virtual reality based simulation of hysteroscopic interventions. Presence: Teleoperators Virtual Environ. **17**(5), 441–462 (2008)

37. Hayward, V., Astley, O.: Performance measures for haptic interfaces. In: Robotics Research: The 7th International Symposium, pp. 195–207 (1996)

38. Hayward, V., Gregorio, P., Astley, O., Greenish, S., Doyon, M., Lessard, L., Mcdougall, J., Sinclair, I., Boelen, S., Chen, X., Demers, J.-P., Poulin, J., Benguigui, I., Almey, N., Makuc, B., Zhang, X.: Freedom-7: a high fidelity seven axis haptic device with application to surgical training. In: Lecture Notes in Control and Information Science, vol. 232, pp. 445–456. Springer, Berlin (1997)

39. Hellier, D., Samur, E., Passenger, J., Spaelter, U., Frimmel, H., Appleyard, M., Bleuler, H., Ourselin, S.: A modular simulation framework for colonoscopy using a new haptic device. In: Proc. of the 16th Medicine Meets Virtual Reality Conference (MMVR), vol. 132, pp. 165–170 (2008)

40. Hocoma AG: Lokomat®. http://www.hocoma.com/ (2012)

41. Hollerbach, J.M., Hunter, I.W., Ballantyne, J.: A comparative analysis of actuator technologies for robotics. In: The Robotics Review, vol. 2, pp. 299–342. MIT Press, Cambridge (1992)

42. Immersion Corp: TouchSense. http://www.immersion.com/products/touchsense-tactile-feedback/ (2012)

43. Intuitive Surgical: da Vinci Surgical System. http://www.intuitivesurgical.com/ (2012)

44. Kappers, A.M., Koenderink, J.J., Lichtenegger, I.: Haptic identification of curved surfaces. Percept. Psychophys. **56 (1)**, 53–61 (1994)

45. King, C.-H., Culjat, M.O., Franco, M.L., Bisley, J.W., Carman, G.P., Dutson, E.P., Grundfest, W.S.: A multielement tactile feedback system for robot-assisted minimally invasive surgery. IEEE Trans. Haptics **2**(1), 52–56 (2009)

46. Kirkpatrick, A.E., Douglas, S.A.: Application-based evaluation of haptic interfaces. In: Proc. of the 10th Haptic Symposium, p. 32 (2002)

47. Kitagawa, M., Dokko, D., Okamura, A.M., Yuh, D.D.: Effect of sensory substitution on suture-manipulation forces for robotic surgical systems. J. Thorac. Cardiovasc. Surg. **129**(1), 151–158 (2005)

48. Lawrence, D.A., Pao, L.Y., Salada, M.A., Dougherty, A.M.: Quantitative experimental analysis of transparency and stability in haptic interfaces. In: Proc. of ASME Dynamic Systems

and Control Division. DSC, vol. 58, pp. 441–449 (1996)

49. Lawrence, D.A., Pao, L.Y., Dougherty, A.M., Salada, M.A., Pavlou, Y.: Rate-hardness: a new performance metric for haptic interfaces. IEEE Trans. Robot. Autom. **16**(4), 357–371 (2000)

50. Leskovsky, P., Cooke, T., Ernst, M., Harders, M.: Using multidimensional scaling to quantify the fidelity of haptic rendering of deformable objects. In: Proc. of Eurohaptics, pp. 289–295 (2006)

51. Logitech: Driving Force™ GT. http://www.logitech.com/en-us/gaming/wheels (2012)

52. Massie, T.H., Salisbury, J.K.: The PHANTOM haptic interface: a device for probing virtual objects. In: Proc. of the ASME Winter Annual Meeting, Symposium on Haptic Interfaces for Virtual Environment and Teleoperator Systems (1994)

53. McAffee, D.A., Fiorini, P.: Hand controller design requirements and performance issues in telerobotics. In: Fifth International Conference on Advanced Robotics, ICAR, vol. 1, pp. 186–192 (1991)

54. McMahan, W., Gewirtz, J., Standish, D., Martin, P., Kunkel, J.A., Lilavois, M., Wedmid, A., Lee, D.I., Kuchenbecker, K.J.: Tool contact acceleration feedback for telerobotic surgery. IEEE Trans. Haptics **4**(3), 210–220 (2011)

55. Mentice SA (formerly Xitact SA): Xitact IHP. http://www.mentice.com/ (2012)

56. Mimic Technologies: dV-Trainer. http://www.mimictech.net/ (2012)

57. MOOG: HapticMaster. http://www.moog.com/products/haptics-robotics/ (2012)

58. Morrell, J.B., Salisbury, J.K.: Parallel-coupled micro-macro actuators. Int. J. Robot. Res. **17**, 773–791 (1998)

59. MPB Technologies: Freedom7. http://www.mpb-technologies.ca/ (2012)

60. Murray, A.M., Klatzky, R.L., Khosla, P.K.: Psychophysical characterization and testbed validation of a wearable vibrotactile glove for telemanipulation. Presence: Teleoperators Virtual Environ. **12**(2), 156–182 (2003)

61. Nintendo: Wii Remote. http://www.nintendo.com/wii/ (2012)

62. Novint Technologies, Inc.: Novint Falcon®

63. Oakley, I., McGee, M.R., Brewster, S.A., Gray, P.D.: Putting the feel in 'look and feel'. CHI, pp. 415–422 (2000)

64. O'Malley, M., Goldfarb, M.: The effect of force saturation on the haptic perception of detail. IEEE/ASME Trans. Mechatron. **7**, 280–288 (2002)

65. Peer, A., Buss, M.: A new admittance-type haptic interface for bimanual manipulations. IEEE/ASME Trans. Mechatron. **13** (2008)

66. Quanser: HD². http://www.quanser.com/ (2012)

67. Ruffaldi, E., Morris, D., Edmunds, T., Barbagli, F., Pai, D.K.: Standardized evaluation of haptic rendering systems. In: Proc. of IEEE Haptic Symposium, pp. 225–232 (2006)

68. Salisbury, C., Gillespie, R.B., Tan, H., Barbagli, F., Salisbury, J.K.: Effects of haptic device attributes on vibration detection thresholds. In: Proc. of World Haptics'09, pp. 115–120 (2009)

69. Salisbury, C.M., Gillespie, R.B., Tan, H.Z., Barbagli, F., Salisbury, J.K.: What you can't feel won't hurt you: evaluating haptic hardware using a haptic contrast sensitivity function. IEEE Trans. Haptics **4**(2), 134–146 (2011)

70. Samur, E., Flaction, L., Bleuler, H.: Design and evaluation of a novel haptic interface for endoscopic simulation. IEEE Trans. Haptics (2011). doi:10.1109/TOH.2011.70

71. Samur, E., Santos-Carreras, L., Sengul, A., Rognini, G., Marchesotti, S., Bleuler, H.: Role of haptics in surgical robotics: report on a workshop. http://www.computer.org/portal/web/toh (2011)

72. Santos-Carreras, L., Beira, R., Sengul, A., Gassert, R., Bleuler, H.: Influence of force and torque feedback on operator performance in a vr-based suturing task. Appl. Bionics Biomech. **7**(3), 217–238 (2010)

73. Semere, W., Kitagawa, M., Okamura, A.M.: Teleoperation with sensor/actuator asymmetry: task performance with partial force feedback. In: Proceedings of the 12th International Conference on Haptic Interfaces for Virtual Environment and Teleoperator Systems, HAPTICS'04, pp. 121–127. IEEE Computer Society, Washington (2004)

74. Sensable Technologies, Inc.®: PHANTOM Omni®. http://www.sensable.com/ (2012)

75. Simbionix USA Corporation: GI-BRONCHI Mentor. http://www.simbionix.com/ (2012)
76. Spaelter, U.: Haptic interface design and control with application to surgery simulation. PhD thesis, EPFL, No. 3529 (2006)
77. Stocco, L.J., Salcudean, S.E., Sassani, F.: Optimal kinematic design of a haptic pen. IEEE/ASME Trans. Mechatron. **6**(3), 210–220 (2001)
78. Surgical Science Sweden AB: LapSim. http://www.surgical-science.com/ (2012)
79. Synaptics Inc.: Synaptics Fuse. http://www.synaptics.com/demos/fuse (2012)
80. Taati, B., Tahmasebi, A.M., Hashtrudi-Zaad, K.: Experimental identification and analysis of the dynamics of a PHANToM premium 1.5A haptic device. Presence: Teleoperators Virtual Environ. **17**(4), 327–343 (2008)
81. Tan, H.: Identification of sphere size using the phantom: towards a set of building blocks for rendering haptic environment. In: ASME Annual Meeting, pp. 197–203 (1997)
82. Tobergte, A., Passig, G., Kuebler, B., Seibold, U., Hagn, U.A., Froehlich, F.A., Konietschke, R., Joerg, S., Nickl, M., Thielmann, S., Haslinger, R., Groeger, M., Nothhelfer, A., Le-Tien, L., Gruber, R., Albu-Schaeffer, A., Hirzinger, G.: MiroSurge—advanced user interaction modalities in minimally invasive robotic surgery. Presence: Teleoperators Virtual Environ. **19**(5, SI), 400–414 (2010)
83. Ueberle, M., Mock, N., Buss, M.: Vishard10, a novel hyper-redundant haptic interface. In: Proceedings of the 12th International Symposium on Haptic Interfaces for Virtual Environment and Teleoperator Systems, HAPTICS'04, pp. 58–65 (2004)
84. Ueberle, M.W.: Design, control, and evaluation of a family of kinesthetic haptic interfaces. PhD thesis, Technische Universität München (2006)
85. Unger, B.J., Nicolaidis, A., Berkelman, P.J., Thompson, A., Klatzky, R.L., Hollis, R.L.: Comparison of 3-d haptic peg-in-hole tasks in real and virtual environments. In: IEEE/RSJ, IROS, pp. 1751–1756 (2001)
86. Van der Linde, R.Q., Lammertse, P., Frederiksen, E., Ruiter, B.: The hapticmaster, a new high-performance haptic interface. In: Proc. of Eurohaptics'02 (2002)
87. Veneman, J.F., Ekkelenkamp, R., Kruidhof, R., van der Helm, F.C.T., van der Kooij, H.: A series elastic- and Bowden-cable-based actuation system for use as torque actuator in exoskeleton-type robots. Int. J. Robot. Res. **25**, 261–281 (2006)
88. VirtaMed: HystSim. http://www.virtamed.com/ (2012)
89. Vollenweider, M.: High quality virtual reality system with haptic feedback. PhD thesis, EPFL, No. 2251 (2000)
90. Wagner, C.R., Stylopoulos, N., Jackson, P.G., Howe, R.D.: The benefit of force feedback in surgery: examination of blunt dissection. Presence: Teleoperators Virtual Environ. **16**(3), 252–262 (2007)
91. Wall, S.A., Harwin, W.: A high bandwidth interface for haptic human computer interaction. Mechatronics **11**, 371–387 (2001)
92. Wall, S.A., Harwin, W.S.: Quantification of the effects of haptic feedback during a motor skills task in a simulated environment. In: Proc. of the 2nd PHANToM Users Research Symposium, pp. 61–69 (2000)
93. Weir, D.W., Colgate, J.E., Peshkin, M.A.: Measuring and increasing z-width with active electrical damping. In: Proc. of IEEE International Symposium on Haptic Interfaces for Virtual Environment and Teleoperator Systems, pp. 169–175 (2008)
94. Weisenberger, J., Kreier, M., Rinker, M.: Judging the orientation of sinusoidal and square-wave virtual gratings presented via 2-dof and 3-dof haptic interfaces. Haptics-e **1**(4) (2000)
95. Yoon, J., Ryu, J.: Design, fabrication, and evaluation of a new haptic device using a parallel mechanism. IEEE/ASME Trans. Mechatron. **6**(3), 221–233 (2001)
96. Zinn, M., Khatib, O., Roth, B., Salisbury, J.K.: Large workspace haptic devices—a new actuation approach. In: Proceedings of the 2008 Symposium on Haptic Interfaces for Virtual Environment and Teleoperator Systems, HAPTICS'08, pp. 185–192. IEEE Computer Society, Washington (2008). doi:10.1109/HAPTICS.2008.4479941

Chapter 3
Human Haptic Perception

Abstract Human hand performs sensory tasks and motor activities at the same time. Highly rich and multi-modal sensory pathway of haptics makes this sensorimotor continuum reliable. Understanding how this continuum works is crucial for any haptics researcher and/or designer. Therefore, this chapter reviews the psychophysical studies on human haptic perception in order to examine capabilities of this continuum. Basic properties of the hand and finger and their limitations in terms of haptic perception are summarized in order to use them as a reference in the following chapters. In addition, common psychophysical testing methods are discussed to provide the necessary background information.

3.1 Human Hand and Haptics

Although haptics is not generally considered a strong pathway for communication of structured information such as text, it is in fact an extremely rich, multi-modal pathway. Some evidence of this lies in the remarkable ability of people to identify common objects quickly and accurately through touch alone [41]. This capability is subserved by a set of stereotyped "Exploratory Procedures" (EPs) such as lateral motion, static contact, pressure, and contour following. Understanding EPs and human tactual identification strategies and taking advantage of them are important steps towards the goal of this work.

A conceptual framework for the human hand function has been introduced by Jones and Lederman [30] which covers tasks that are sensory in nature to ones involve motor activities. This sensorimotor continuum is divided into four parts. *Tactile sensing* includes contact between the stationary hand (i.e. passive) and an object. The object may not necessarily be stationary. On the other hand, *active haptic sensing* defines contact of the voluntarily moving hand over an object and dominated for identifying an object. Two modes closer to the human motor system are *prehension* which involves basic tasks to grasp an object and *non-prehensile skilled movements* which are related to the gestures of the hand. This sensorimotor continuum defines the base for haptic interaction tasks.

A model of the human haptic system in relation to object identification proposed by Klatzky and Lederman [33] suggests that tangible object properties can be categorized into two: material properties (texture, roughness, hardness, thermal and weight) and structural or geometrical cues (shape, size, orientation, curvature, patterns). This distinction has also been supported by a recent fMRI study [52]. Moreover, psychophysical studies on human haptic identification of objects show that material properties compensate haptic object recognition in case of limited access to object structural cues (shape and size) but are not the primary basis for haptic object identification [34]. In other words, objects can be recognized by geometrical cues.[1]

Understanding the mechanical properties of hand and finger is also crucial to examine its capabilities. Hajian and Howe [15] investigated the mechanical impedance at the human finger tip for the force range of 2–20 N for extension and 2–8 N for abduction. Estimated finger tip mass for each subject was constant for force levels greater than 6 N for extension and constant for all the abduction trials. The estimated stiffness increased linearly from 200 N/m to 800 N/m and damping from 2.2 Ns/m to 4.0 Ns/m with extension muscle activation. Apart from the increase of joint stiffness or damping due to muscle activation, cocontraction of antagonistic muscle groups is also used to increase the joint impedance and natural frequency [48, 68]. For instance, experiments performed by Milner and Cloutier [49] to find the impedance characteristics of the wrist joint showed that stiffness changes from 2 Nm/rad to 6 Nm/rad and damping from 0.04 Nms/rad to 0.08 Nms/rad with muscle coactivation. They also showed that coactivation is used to raise the natural frequency of the wrist (which is around 7–8 Hz) to guarantee stability of the joint.

The response of the finger pad under dynamic compressions is characterized by Serina et al. [58] to understand the force modulation by the fingertip. They found a nonlinear force-displacement relationship. Displacement mostly occurs for forces of a small magnitude (<1 N corresponding up to 2–2.5 mm displacement of the finger pad) while higher forces are transmitted to the bone. This suggests that the fingertip functions effectively as a tactile sensor within small force ranges because small force changes result in enhancement of mechanoreceptor activity related to displacement and contact area [30]. At this level of force, the contact area is around 150 mm^2 [73].

Nakazawa et al. [50] described impedance characteristics of the human fingertips in the tangential directions by a viscoelastic Kelvin model. They showed that while a constant normal force of 1.9 N is applied to the fingertip, stiffness and damping parameters decrease as shear force increases. However, these constants stabilize (stiffness of 400 N/m and damping of 20 Ns/m) for shear forces higher than 1.7 N.

3.2 Psychophysics and Psychophysical Methods

Studies to reveal how the hand and fingers function can be categorized into two fields: neurophysiology and psychophysics. The definitions of these fields given by Jones and Lederman [30] are as follows:

---

[1]In the text, geometrical and force cues are used in a correlated way. Otherwise, shape features are identified by force cues [56].

Neurophysiology the study of the relation between physical stimuli (physical events) and responses of single afferent units (neural events).

Psychophysics the study of determining the mathematical function describing the relation between physical stimuli (physical events) and human tactual sensory experiences (mental events).

Although the neurophysiology provides insightful information on how the mechanoreceptors work in the hand, only psychophysical findings on human haptic perception are summarized in this chapter. Before presenting these findings, it is as well useful to look at the experimental methods used for psychophysical analysis in general.

There are three general methods in psychophysics that are used to determine human thresholds of perception: *detection, discrimination*, and *identification*. Detection and discrimination involve the measurement of sensory thresholds of perception of a stimulus. Identification, on the other hand, involves human ability to categorize stimuli without providing explicit references [63]. Psychophysical measurement of thresholds results in two estimations:

Absolute threshold is the minimum amount of stimulation required for a human to detect a stimulus. In other words, it represents the human ability to perceive the smallest intensity that is just detectable. "RL" is used as the acronym of the absolute threshold which originates from the German *Reiz Limen* [12].

Differential threshold is the smallest difference in a stimulus detectable by a human. A discrimination test results in a differential threshold. The differential threshold is also known as the *just-noticeable-difference* (JND) or *difference threshold* or *difference limen* (DL). Since the differential threshold is measured relative to the stimulus, it should be presented with respect to the base value. Alternatively, it is represented by a ratio called *Weber fraction*:

$$c = \frac{\Delta I}{I} \tag{3.1}$$

where ΔI is the differential threshold and I is the magnitude of the reference stimulus. This dimensionless representation generally results in a constant value for sensory threshold. This is called *Weber's law*.

These two thresholds can be measured by similar experimental tests. The standard and most utilized experimental methods for sensory threshold estimations are described below. The complete list and details can be found in [12] and [46].

1. Method of constant stimuli: It is the most preferred standard experimental method to measure absolute and differential thresholds. Equally spaced stimuli (5 to 9 values) are presented in a random order as shown in Fig. 3.1. This method reduces expectation errors resulting from human prediction of certain scenarios, occurring in case of presenting stimuli in ascending or descending order. Although it provides accurate results, it is not efficient in terms of time consumption.

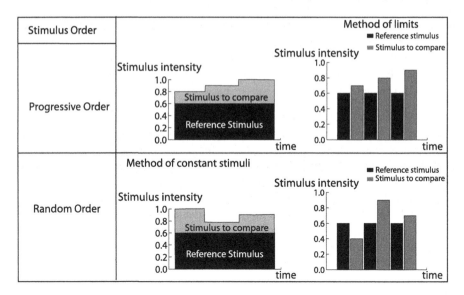

Fig. 3.1 Stimulus presentation to the subject for two methods: the method of constant stimuli and the method of limits

For the absolute threshold, the subject responses whether the given stimulus is detectable or not, by saying "yes" or "no". For each stimulus value, the proportions of "yes" responses is calculated. These proportions can be drawn against the stimuli intensity to obtain a graph called *psychometric function*. The absolute threshold is found by setting a threshold to the proportions at 50 %. It means that this intensity would be detected 50 % of the time.

In order to measure differential thresholds using the method of constant stimuli; two stimuli, one of which being the reference, are ideally presented simultaneously. Subjects are asked to indicate whether the other stimulus is perceived differently than the reference stimulus or not. 5 to 9 values are required in order to obtain the psychometric function. 75 % "different" answer is taken as the differential threshold. There are two practical problems of this method. Ideally, both stimuli should be presented together in space and time which is not possible in practice.

2. Method of limits: It is less precise than the method of constant stimuli but more efficient. Expectation and habituation are two sources of errors. Ascending or descending stimuli (see Fig. 3.1) are given well below or above the threshold until the subject reports presence or absence. This point is taken as the absolute threshold. In order to calculate the differential threshold, the interval where the subject cannot perceive any difference is taken. Half of this interval is taken as the differential threshold.

There are two basic variations of this method: Staircase (up-down) and forced choice [43]. In the staircase method, when the response of the subject changes (i.e., "no" to "yes"), the ascending order changes to descending, or vice versa.

This continues until a sufficient number of response are obtained (i.e., 75 %). It is a very fast method but suffers from the improper size of steps. On the other hand, the forced-choice method requires subjects to choose the stimulus which actually contains the stimulus or is greater. This reduces the inherent errors of the method of limits.

Another variation of the method of limits is proposed by Kaernbach [32]. In this method, an adaptive staircase method is used with unforced-choice. Additional response option of "don't know" is given to the subject which makes them more comfortable during the trials. In addition, the step size for the next trial is adjusted according to the subject's response. The step sizes after correct, incorrect and unsure answers, which are designated as S_{cor}, S_{incor} and S_{unsure} respectively, are calculated based on the target performance. If 75 % is chosen to be the target, the step sizes for a two alternative discrimination test are

$$S_{incor} = -3 \times S_{cor} \tag{3.2}$$

$$S_{unsure} = -S_{cor} \tag{3.3}$$

3. Method of adjustments: In this method, the stimulus is set to an arbitrary value and subjects are asked to adjust the intensity level until he/she reports presence or absence of the stimulus or the difference. It is not as accurate as the others, therefore it is only recommended for preliminary experiments to predict the results.

4. Magnitude estimation: Subjects are required to make direct numerical estimation of the stimuli (10 to 20 stimuli). Either a standard stimuli is given to the subject with a certain numerical value as a reference, or the subject is left free to assign any number to his sensations that best matches the perceived magnitude of the stimuli. Data from several subjects are collected and the median or geometric mean is calculated. The subjective magnitude estimation (or perceived magnitude of stimulus) is drawn against the real stimulus value. A good example for this method is Lederman's tactile roughness experiments where the perceived roughness is plotted against the real groove width [40]. The method is a fast and quantitative way of scaling sensations. Therefore, results are generally expressed as "accurately scaled" or "capable of accurately scaling" over the stimuli range.

3.3 Human Haptic Perception Limits

In order to use the human perception as an evaluation tool for haptic interfaces, we have to understand the limits and capabilities of the human hand. For this reason, the psychophysical experiments on the human hand sensory system were surveyed. The findings are summarized in four tables: Table 3.1 shows the parameters related to perception of limb motion and position (i.e., *kinesthesia*), Table 3.2 focuses on the perception thresholds for a variety of stimuli, Table 3.3 summarizes how much detail a human can perceive from an object and finally Table 3.4 shows the output capabilities of a human hand. In the tables, references to the original papers are also given. For a complete review of human haptic perception, one can consult [10, 30].

Table 3.1 Kinesthesia: Perception of limb motion and position. Unless otherwise stated, the values are for the index finger

Parameter		Value	Notes	Reference
Position threshold	Absolute	0.6–1.1°	Elbow & shoulder	[69]
	Differential	1.7–2.7°		[66]
		2.0/0.8°	Wrist & elbow/shoulder	[64]
		9 %	Finger & elbow	[26, 28]
Movement threshold	Absolute	1°	For the velocity range from 10°/s to 80°/s	[16]
		0.2–0.7°	Wrist, elbow & shoulder	[6]
	Differential	8 %	Elbow	[31]
Information transfer		1.92 bits	Identification of 3–4 joint angle positions	[66]
Sensing bandwidth		20–30 Hz	Proprioceptive & kinesthetic sensors	[3]

Table 3.2 Sensory thresholds of human hand. Unless otherwise stated, the values are for the index fingertip

Parameter		Value	Notes	Reference
Force	Absolute	0.06 N		[26]
	Differential	7 %	Over a range of 2.5–10 N	[53]
		7 %	Elbow muscle. Over a range of 25–400 N	[25, 64]
		15–27 %	For forces less than 0.5 N	[26]
Pressure	Absolute	0.019 g	Women	[30]
		0.055 g	Men	
Friction	Accurately scaled	0.43–2.79	Range of friction coefficient	[59]
Shear	Accurately scaled	0.15–0.70 N		[54]
Vibration	Intensity (differential)	25 %	Over a range of 10–20 dB sensation level	[8]
	Temporal (absolute)	const 28 dB	For 0.4–3 Hz with 1 μm peak	[1]
		−5 dB/dec	For 3–30 Hz	[70]
		−10 dB/dec	For 30–250 Hz	
		+10 dB/dec	For 250–700 Hz	
Spatial resolution		1.2 mm	Discrimination of grating orientation	[9, 23]
		0.6 mm	Detection of grating	[13]
Temporal resolution		5 ms	Successiveness of mechanical pulses	[30]

Table 3.3 Perception of object properties

Parameter		Value	Notes	Reference
Roughness	Differential threshold	% 2	Using 2-D raised dots	[37]
	Model		Spatial intensive. Determined primarily by cutaneous inputs	[7, 40]
	Period	1.5–8.5 mm	Magnitude estimates are linear	[60]
	Maximum perceived roughness	5.0 mm	Groove width. Bare finger	[39]
		3.5 mm	Groove width. Through a rigid probe of 3 mm diameter	[35]
Texture	Orthogonal directions		Rough/smooth, hard/soft, sticky/slippery	[20, 21]
Micro-geometry	Detection threshold	2 μm	Height of a single dot. Passive detection	[38]
		0.06 μm	Height of periodically arranged bars. Detectable while active movement	
		0.16 μm	Height of periodically arranged dots. Detectable while active movement	
Shape	Length discrimination	1 mm	For 10–20 mm	[11]
		2.5 mm	For 80 mm	[67]
	Thickness discrimination	0.075 mm	For 0.2 mm	[22]
				[19]

Table 3.3 (continued)

Parameter		Value	Notes	Reference
	Curvature detection	1 μm to 8 mm	Depending on width of the profile. Active	[45]
		0.56 m^{-1}	Active	[18]
	Curvature discrimination	2.26 m^{-1}	Active	[18]
	Grating discrimination	5 %	Spatial period of 0.77–1.0 mm. Active	[14]
		10 %	Spatial period of 0.77–1.0 mm. Passive	
	Angle discrimination	5°	Range 0.7 to 12.1°	[71]
Material	Mass discrimination	21 %	Reference mass of 12 kg	[42]
	Compliance discrimination	5–15 %		[65]
				[62]
	Stiffness discrimination	17 %		[30]
		23 %		[28]
	Viscosity discrimination	34 %	Viscosity matching using full arm	[29]
		14 %	Viscosity discrimination using two fingers	[42]
	Weight discrimination	8 %	Decreases to 6 % by jiggling	[2, 24]
	Rigidity	25 N/mm	Object is felt rigid at this value	[64]
Recognition Time	Initial haptic glance	200 ms		[30]
	Response latency	1–2 s		

Table 3.4 Output capabilities of human

Parameter		Value	Notes	Reference
Force	Maximum	50/60 N	Finger/wrist	[64]
		100 N	Elbow & shoulder	
	Control	11–15 %		[61]
		1 %	Over 10 to 20 N	[64]
		1 %	Wrist, elbow & shoulder Over 20 to 50 N	
		10 %	Over 5 to 18 N	[47]
	Bandwidth	2–3 Hz		[55]
		5 Hz		[3]
		2–6 Hz		[26, 27]
Motion	Maximum speed	17.6 rad/s		[17]
	Bandwidth	<2 Hz	Active touch for sensing	[36]
		2–4 Hz	Voluntary movements	[44, 51]
		2–7 Hz	Periodic tracking	[5]
		4–8 Hz	Skilled actions: hand writing, typing, tapping, playing musical instruments	[26, 36]
		10 Hz	Reflexive actions	[3, 4]
		<10 Hz		[72]
		8–12 Hz	Finger tremor	[57]

References

1. Bolanowski, S.J., Gescheider, G.A., Verrillo, R.T., Checkosky, C.M.: Four channels mediate the mechanical aspects of touch. J. Acoust. Soc. Am. **84**(5), 1680–1694 (1988)
2. Brodie, E.E., Ross, H.E.: Jiggling a lifted weight does aid discrimination. Am. J. Psychol. **98**(3), 469–471 (1985)
3. Brooks, T.L.: Telerobotic response requirements. In: IEEE International Conference on Systems, Man and Cybernetics, pp. 113–120 (1990)
4. Carter, R.R., Crago, P.E., Keith, M.W.: Stiffness regulation by reflex action in the normal human hand. J. Neurophysiol. **64**(1), 105–118 (1990)
5. Cathers, I., O'Dwyer, N., Neilson, P.: Tracking performance with sinusoidal and irregular targets under different conditions of peripheral feedback. Exp. Brain Res. **111**(3), 437–446 (1996)
6. Clark, F.J., Horch, K.W.: Kinesthesia. In: Boff, K.R., Kaufman, L., Thomas, J.P. (eds.) Handbook of Perception and Human Performance: Sensory Processes and Perception, vol. 1 (1986)
7. Connor, C.E., Johnson, K.O.: Neural coding of tactile texture: comparison of spatial and temporal mechanisms for roughness perception. J. Neurosci. **12**(9), 3414–3426 (1992)
8. Craig, J.C.: Vibrotactile difference thresholds for intensity and the effect of a masking stimulus. Percept. Psychophys. **15**(1), 123–127 (1974)
9. Craig, J.C.: Grating orientation as a measure of tactile spatial acuity. Somatosens. Motor Res. **16**(3), 197–206 (1999)

10. Durlach, N.I., Mavor, A.S.: Virtual Reality: Scientific and Technological Challenges. National Academy Press, Washington (1995)

11. Durlach, N.I., Delhorne, L.A., Wong, A., Ko, W.Y., Rabinowitz, W.M., Hollerbach, J.: Manual discrimination and identification of length by the finger-span method. Percept. Psychophys. **46**(1), 29–38 (1989)

12. Gescheider, G.A.: Psychophysics: The Fundamentals, 3rd edn. Lawrence Erlbaum Associates, Mahwah (1997)

13. Gibson, G.O., Craig, J.C.: Relative roles of spatial and intensive cues in the discrimination of spatial tactile stimuli. Percept. Psychophys. **64**(7), 1095–1107 (2002)

14. Morley, J.W., Goodwin, A.W., Darian-Smith, I.: Tactile discrimination of gratings. Exp. Brain Res. **49**(2), 291–299 (1983). doi:10.1007/BF00238588

15. Hajian, A.Z., Howe, R.D.: Identification of the mechanical impedance at the human finger tip. J. Biomech. Eng. **119**(1), 109–114 (1997)

16. Hall, L.A., McCloskey, D.I.: Detections of movements imposed on finger, elbow and shoulder joints. J. Physiol. **335**, 519–553 (1983)

17. Hasser, C.J.: Force-reflecting anthropomorphic hand masters. Armstrong Labortory Technical Report (1995)

18. Henriques, D.Y.P., Soechting, J.F.: Bias and sensitivity in the haptic perception of geometry. Exp. Brain Res. **150**(1), 95–108 (2003)

19. Ho, C., Srinivasan, M.A.: Human haptic discrimination of thickness. Touch Lab Report, RLE TR-608, 6 (1997)

20. Hollins, M., Faldowski, R., Rao, S., Young, F.: Perceptual dimensions of tactile surface texture: a multidimensional scaling analysis. Percept. Psychophys. **54**(6), 697–705 (1993)

21. Hollins, M., Bensmaia, S., Karlof, K., Young, F.: Individual differences in perceptual space for tactile textures: evidence from multidimensional scaling. Percept. Psychophys. **62**(8), 1534–1544 (2000)

22. John, K.T., Goodwin, A.W., Darian-Smith, I.: Tactile discrimination of thickness. Exp. Brain Res. **78**(1), 62–68 (1989)

23. Johnson, K.O., Phillips, J.R.: Tactile spatial resolution. i. Two-point discrimination, gap detection, grating resolution, and letter recognition. J. Neurophysiol. **46**(6), 1177–1191 (1981)

24. Jones, L.A.: Perception of force and weight. Theory and research. Psychol. Bull. **100**(1), 29–42 (1986)

25. Jones, L.A.: Matching forces: constant errors and differential thresholds. Perception **18**(5), 681–687 (1989)

26. Jones, L.A.: Kinesthetic sensing. In: Human and Machine Haptics. MIT Press, Cambridge (2000, Unpublished manuscript)

27. Jones, L.A., Hunter, I.W.: Influence of the mechanical properties of a manipulandum on human operator dynamics. 1 Elastic stiffness. Biol. Cybern. **62**(4), 299–307 (1990)

28. Jones, L.A., Hunter, I.W.: A perceptual analysis of stiffness. Exp. Brain Res. **79**(1), 150–156 (1990)

29. Jones, L.A., Hunter, I.W.: A perceptual analysis of viscosity. Exp. Brain Res. **94**, 343–351 (1993)

30. Jones, L.A., Lederman, S.J.: Human Hand Function. Oxford University Press, London (2006)

31. Jones, L.A., Hunter, I.W., Irwin, R.J.: Differential thresholds for limb movement measured using adaptive techniques. Percept. Psychophys. **52**(5), 529–535 (1992)

32. Kaernbach, C.: Adaptive threshold estimation with unforced-choice tasks. Percept. Psychophys. **63**(8), 1377–1388 (2001)

33. Klatzky, R.L., Lederman, S.J.: Toward a computational model of constraint-driven exploration and haptic object identification. Perception **22**(5), 597–621 (1993)

34. Klatzky, R.L., Loomis, J.M., Lederman, S.J., Wake, H., Fujita, N.: Haptic identification of objects and their depictions. Percept. Psychophys. **54 (2)**, 170–178 (1993)
35. Klatzky, R.L., Lederman, S.J., Hamilton, C., Grindley, M., Swendsen, R.H.: Feeling textures through a probe: effects of probe and surface geometry and exploratory factors. Percept. Psychophys. **65**(4), 613–631 (2003)
36. Kunesch, E., Binkofski, F., Freund, H.-J.: Invariant temporal characteristics of manipulative hand movements. Exp. Brain Res. **78**(3), 539–546 (1989)
37. Lamb, G.D.: Tactile discrimination of textured surfaces: psychophysical performance measurements in humans. J. Physiol. **338**(1), 551–565 (1983)
38. LaMotte, R.H., Srinivasan, M.A.: Surface microgeometry: tactile perception and neural encoding. In: Information Processing in the Somatosensory System, pp. 49–58. Macmillan Co, New York (1991)
39. Lawrence, M.A., Kitada, R., Klatzky, R.L., Lederman, S.J.: Haptic roughness perception of linear gratings via bare finger or rigid probe. Perception **36**(4), 547–557 (2007)
40. Lederman, S.J.: Tactile roughness of grooved surfaces: the touching process and effects of macro- and microsurface structure. Percept. Psychophys. **16**(2), 385–395 (1974)
41. Lederman, S.J., Klatzky, R.L.: Hand movements: a window into haptic object recognition. Cogn. Psychol. **19**, 342–368 (1987)
42. Lee Beauregard, G., Srinivasan, M.A., Durlach, N.I.: Manual resolution of viscosity and mass, vol. 57-2, pp. 657–662 (1995)
43. Leek, M.R.: Adaptive procedures in psychophysical research. Percept. Psychophys. **63**(8), 1279–1292 (2001)
44. Leist, A., Freund, H.-J., Cohen, B.: Comparative characteristics of predictive eye-hand tracking. Hum. Neurobiol. **6**(1), 19–26 (1987)
45. Louw, S., Kappers, A.M.L., Koenderink, J.J.: Haptic detection thresholds of Gaussian profiles over the whole range of spatial scales. Exp. Brain Res. **132**(3), 369–374 (2000)
46. Macmillan, N.A., Creelman, C.D.: Detection Theory: A User's Guide, 2nd edn. Psychology Press/Taylor & Francis, New York/London (2008)
47. Mai, N., Schreiber, P., Hermsdorfer, J.: Changes in perceived finger force produced by muscular contractions under isometric and anisometric conditions. Exp. Brain Res. **84**(2), 453–460 (1991)
48. Milner, T.E.: Adaptation to destabilizing dynamics by means of muscle cocontraction. Exp. Brain Res. **143**(4), 406–416 (2002)
49. Milner, T.E., Cloutier, C.: Damping of the wrist joint during voluntary movement. Exp. Brain Res. **122**(3), 309–317 (1998)
50. Nakazawa, N., Ikeura, R., Inooka, H.: Characteristics of human fingertips in the shearing direction. Biol. Cybern. **82**(3), 207–214 (2000)
51. Neilson, P.D.: Speed of response or bandwidth of voluntary system controlling elbow position in intact man. Med. Biol. Eng. **10**(4), 450–459 (1972)
52. Newman, S.D., Klatzky, R.L., Lederman, S.J., Just, M.A.: Imagining material versus geometric properties of objects: an fmri study. Cogn. Brain Res. **23**(2–3), 235–246 (2005)
53. Pang, X.D., Tan, H.Z., Durlach, N.I.: Manual discrimination of force using active finger motion. Percept. Psychophys. **49**(6), 531–540 (1991)
54. Paré, M., Carnahan, H., Smith, A.M.: Magnitude estimation of tangential force applied to the fingerpad. Exp. Brain Res. **142**(3), 342–348 (2002)
55. Partridge, L.D.: Signal-handling characteristics of load-moving skeletal muscle. Am. J. Physiol. **210**(5), 1178–1191 (1966)
56. Robles-De-La-Torre, G., Hayward, V.: Force can overcome object geometry in the perception of shape through active touch. Nature **412**, 445–448 (2001)
57. Safwat, B., Su, E.L.M., Gassert, R., Teo, C.L., Burdet, E.: The role of posture, magnification, and grip force on microscopic accuracy. Ann. Biomed. Eng. **37**(5), 997–1006 (2009)
58. Serina, E.R., Mote, C.D., Rempel, D.: Force response of the fingertip pulp to repeated compression—effects of loading rate, loading angle and anthropometry. J. Biomech. **30**(10), 1035–1040 (1997)

59. Smith, A.M., Scott, S.H.: Subjective scaling of smooth surface friction. J. Neurophysiol. **75**(5), 1957–1962 (1996)
60. Smith, A.M., Chapman, C.E., Deslandes, M., Langlais, J.-S., Thibodeau, M.-P.: Role of friction and tangential force variation in the subjective scaling of tactile roughness. Exp. Brain Res. **144**(2), 211–223 (2002)
61. Srinivasan, M.A., Chen, J.S.: Human performance in controlling normal forces of contact with rigid objects. In: ASME, Advances in Robotics, Mechatronics, and Haptic Interfaces, DSC, vol. 49 (1993)
62. Srinivasan, M.A., LaMotte, R.H.: Tactual discrimination of softness. J. Neurophysiol. **73**(1), 88–101 (1995)
63. Tan, H.: Identification of sphere size using the phantom: towards a set of building blocks for rendering haptic environment. In: ASME Annual Meeting, pp. 197–203 (1997)
64. Tan, H.Z., Eberman, B., Srinivasan, M.A., Cheng, B.: Human factors for the design of force-reflecting haptic interfaces. ASME Dyn. Syst. Control **55**(1), 353–359 (1994)
65. Tan, H.Z., Durlach, N.I., Beauregard, G.L., Srinivasan, M.: Manual discrimination of compliance using active pinch grasp: the roles of force and work cues. Percept. Psychophys. **57**(4), 495–551 (1995)
66. Tan, H.Z., Srinivasan, M.A., Reed, C.M., Durlach, N.I.: Discrimination and identification of finger joint-angle position using active motion. ACM Trans. Appl. Percept. **4**(2), 10 (2007)
67. Tan, H.Z., Pang, X.-D., Durlach, N.I.: Manual resolution of length, force, and compliance
68. Tee, K.P., Franklin, D.W., Kawato, M., Milner, T.E., Burdet, E.: Concurrent adaptation of force and impedance in the redundant muscle system. Biol. Cybern. **102**(1), 31–44 (2010)
69. VanBeers, R.J., Sittig, A.C., Denier VanDerGon, J.J.: The precision of proprioceptive position sense. Exp. Brain Res. **122**(4), 367–377 (1998)
70. Verrillo, R.T.: Effect of contactor area on the vibrotactile threshold. J. Acoust. Soc. Am. **35**(12), 1962–1966 (1963)
71. Voisin, J., Benoit, G., Chapman, E.C.: Haptic discrimination of object shape in humans: two-dimensional angle discrimination. Exp. Brain Res. **145**(2), 239–250 (2002)
72. Wall, S.A., Harwin, W.: A high bandwidth interface for haptic human computer interaction. Mechatronics **11**, 371–387 (2001)
73. Westling, G., Johansson, R.S.: Factors influencing the force control during precision grip. Exp. Brain Res. **53**(2), 277–284 (1984)

Part II
Physical Evaluation

Chapter 4
Performance Evaluation Based on Physical Measurements

Abstract Although a set of performance measures for haptic interfaces has already been defined in the literature, it is almost impossible to find detailed information on testing conditions and methods. Since there is no consensus on measurement methods, which vary considerably across studies, it might be misleading to compare technical specifications of different devices. In order to obtain a coherent technical evaluation methodology with specific performance metrics, we study physical evaluation methodologies for haptic interfaces in this chapter. Existing physical performance measures and methodologies in the field of robotics and haptics are categorized and described in detail. This results in a tutorial-like guideline for physical device evaluation, which describes in detail testing conditions and methods.

4.1 Introduction

Physical performance evaluation includes device characterization and technical assessment which results in specifications of a device. The discussion about experimental performance evaluation for haptic interfaces goes back to 80s when the design requirements for teleoperation were first described. Since then, basic performance characteristics of haptic interfaces have identified by many researchers. Key studies on experimental performance evaluation for haptic interfaces are given in Table 4.1. In several other projects [1, 4, 12, 14, 27, 30, 35, 37, 39], researchers evaluated their particular haptic devices based on the performance metrics given in these key studies. In the following sections, these physical performance metrics for haptic interfaces are described in detail.

4.2 Modeling Haptic Interface Performance

In order to express haptic interface performance, a model describing haptic interaction is necessary. One suitable way to do it is to use the notion of exchange of energy between the human user, the haptic interface and the virtual environment. This energy exchange is commonly represented by a black box which captures the relationship between forces and velocities [1].

E. Samur, *Performance Metrics for Haptic Interfaces*,   43
Springer Series on Touch and Haptic Systems,
DOI 10.1007/978-1-4471-4225-6_4, © Springer-Verlag London 2012

Table 4.1 Key studies on experimental performance evaluation for haptic interfaces

Authors	Contributions
Eppinger [11]	Described guidelines to measure output force capability of a robot
Brooks [5]	Defined teleoperation design requirements
McAffee and Fiorini [23]	Identified key performance characteristics for teleoperation
Hollerbach et al. [21]	Made a comparative analysis of actuator technologies for robotics
Hayward and Astley [18]	Theoretically defined performance measures for haptic interfaces
Morrell and Salisbury [26]	Demonstrated performance measures for coupled micro-macro actuators
Ellis et al. [10]	Experimentally demonstrated practical ways to measure performance measures on a haptic interface
Colgate and Brown [8]	Introduced the Z-width concept which is the impedance range of a haptic device over frequency
Frisoli and Bergamasco [13]	Described an experimental identification method
Cavusoglu [6]	Experimentally identified the dynamics of PHANTOM Premium 1.5A haptic device
Ueberle [32, 33]	Conducted hardware experiments for performance evaluation of haptic control schemes
Weir et al. [36]	Described methods to measure impedance range of a haptic device
Chapuis [7]	Proposed a method to calculate output impedance of a device

This interaction dynamics may also be characterized by *mechanical impedance*, which may be considered a dynamic extension of stiffness [20]. It is the system's resistance to a motion. A more precise definition of impedance is the complex transfer function between a given velocity input and resulting force output:

$$Z(\omega) = \frac{F(\omega)}{v(\omega)} \tag{4.1}$$

A model of the haptic interaction as an electronic circuit diagram is shown in Fig. 4.1. In this model, only an impedance type of haptic interface and virtual reality is considered. An *impedance haptic interface* measures the velocity of the end effector and responds with the force. In the following sections, this convention is preserved and all the analysis are done for this model. For a model of admittance type interfaces, which measure the force of the end effector and responds with the velocity, reader should refer to [27].

In Fig. 4.1, the human hand is modeled by two components [31]:

1. Effort Source (F_h), i.e., the resulting force from voluntary muscle (co-)contraction as a function of time or frequency.

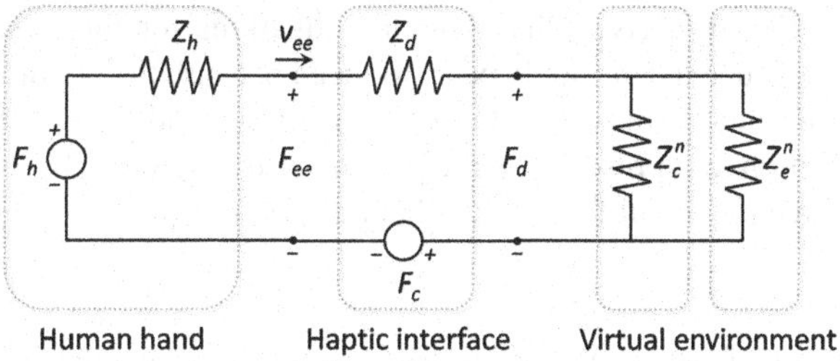

Fig. 4.1 Electronic circuit representation of haptic rendering. The human operator and the haptic interface share the same force (F_{ee}) and velocity (v_{ee}) at the point of interaction (at the end effector). The virtual environment consist of two parts: virtual coupling (Z_c^n) ensuring stable rendering and actual rendered environment (Z_e^n)

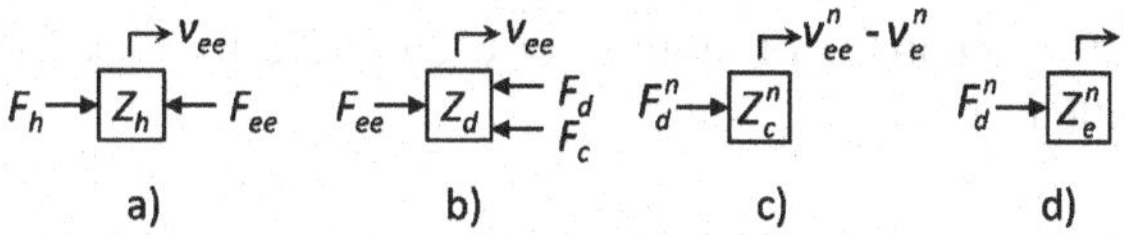

Fig. 4.2 Free body diagrams of components involved in haptic rendering: (**a**) human hand, (**b**) haptic interface, (**c**) virtual coupler and (**d**) virtual environment

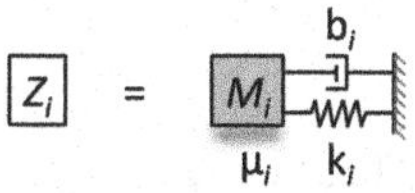

Fig. 4.3 1-DOF mechanical equivalent of an impedance representation. Inertia, damping, stiffness, losses such as friction and all other disturbances are lumped in the impedance

2. Impedance (Z_h), i.e., variable depending on posture and muscle contraction. It is also either a function of time or frequency.

F_{ee} is the force exerted to the haptic interface and also the force felt by the human user at the end effector. Velocity of the end effector is designated by v_{ee} which equals to the human hand velocity. The desired force (F_d) given to the haptic interface is subject to the impedance of the device (Z_d). Therefore, a control effort (F_c) is required to compensate for the device dynamics. The virtual environment is divided into two parts: virtual coupling (Z_c) and actual virtual environment (Z_e). Unlike the notation used in the literature [1, 15], the virtual coupler ensuring stable rendering is considered to be a part of the virtual world.

Mechanical models of haptic rendering in terms of free body diagrams are shown in Fig. 4.2. Here, an impedance (Z_i) is the lumped representation of all dynamic factors such as mass/inertia (M_i), damping (d_i), stiffness (k_i), losses such as friction (μ_i) and all other disturbances. They are represented in the Cartesian space.

Figure 4.3 shows a 1-DOF mechanical equivalent of this impedance representation. In this case, the impedance in the Laplace domain is

$$Z_i(s) = \left(s M_i + b_i + \mu_i + \frac{k_i}{s} \right) \tag{4.2}$$

Before deriving the dynamic equations for this system, an electro-mechanical transducer term should be introduced because the haptic interface itself is an electro-mechanical transducer between the digital and analog world. The superscript "n" is chosen to represent that the parameter is discrete. Then, the relation between output force and input force of the haptic interface is

$$F_d = H_f F_d^n \tag{4.3}$$

where F_d^n is the digital desired force and H_f is the transfer function comprising all signal conversions and kinematic transformations realized by the haptic interface. Similarly, a transfer function for the velocity measurement can be written:

$$v_{ee}^n = H_v v_{ee} \tag{4.4}$$

The equations of motion for each of the free body diagrams in Fig. 4.2 can be written as

$$F_h - F_{ee} = Z_h v_{ee} \tag{4.5}$$

$$F_{ee} - F_d - F_c = Z_d v_{ee} \tag{4.6}$$

$$F_d^n = Z_c^n \left(v_{ee}^n - v_e^n \right) \tag{4.7}$$

$$F_d^n = Z_e^n v_e^n \tag{4.8}$$

These equations provide useful information about the haptic interaction. For example, when the desired force is set to zero in Eq. (4.6), the human feels the dynamics of the device unless there is a control action. Therefore, a controller should be implemented for compensation of device dynamics as

$$F_c^n = -Z_d^n v_{ee}^n \tag{4.9}$$

Here, the performance of the controller depends on how close the real device impedance is represented in the digital domain (i.e. $Z_d^n \simeq Z_d$).

Equations (4.5) and (4.6) can be solved to calculate the two outputs, i.e. force (F_{ee}) and the velocity (v_{ee}) of the end effector. For an uncontrolled haptic interface ($F_c = 0$), this results in

$$\begin{bmatrix} F_h \\ F_{ee} \end{bmatrix} = \begin{bmatrix} Z_h + Z_d & H_f \\ Z_d & H_f \end{bmatrix} \begin{bmatrix} v_{ee} \\ F_d^n \end{bmatrix} \tag{4.10}$$

The first row represents the dynamic equation of the haptic system: the velocity of the end effector is governed by the human induced force and the desired force. On the other hand, the second equation shows that the desired force is degraded only by the device dynamics for this uncontrolled system. Therefore, a feedforward control should be implemented in order to compensate for the device dynamics and to obtain a transparent system:

$$F_{ee} = \left(Z_d - H_f Z_d^n H_v \right) v_{ee} + H_f F_d^n \tag{4.11}$$

Although the human impedance does not play any role in the transparency, the controller performance is highly related to the overall system's stability including the human impedance. If all device dynamics is compensated, an unstable system is obtained. Therefore, the controller performance should be measured when the human is interacting with the haptic interface.

The matrix representation in Eq. (4.10) without the virtual environment and the control is also useful when the device dynamics need to be extracted. At this point, a method is proposed by Chapuis to calculate the output impedance of a device using the electrical analogy [7]. Let us consider two cases:

1. Open end (i.e. free output, not interacting with the human). In this case, F_h, Z_h and hence F_{ee} will be zero. Therefore, the end effector velocity is calculated as

$$v_{ee} = -\frac{H_f}{Z_d} F_d^n = -Y_f F_d^n \qquad (4.12)$$

Here, Y_f can be called the *modified admittance* of the device as the real admittance of the device ($Y_d = Z_d^{-1}$) is magnified by the transfer function H_f.

2. Fixed end (i.e. blocked output, the end effector is constrained). In this case, we can assume that $Z_h \cong \infty$. Therefore, v_{ee} will be zero and the force at the end effector will be

$$F_{ee} = H_f F_d^n \qquad (4.13)$$

This equation allow us to calculate the transfer function (H_f) between input and output of the haptic interface.

Using the electrical analogy and comparing the open end (i.e. open circuit) and fixed end (i.e. short circuit) solutions, the output impedance can be calculated [7]. Dividing Eq. (4.13) by Eq. (4.12) results in:

$$Z_d = \frac{F_{ee}}{v_{ee}} \qquad (4.14)$$

which is the output impedance of the device.

The input-output relation given in Eq. (4.10) can be combined with the virtual environment variables in order to obtain the dynamic equation of the overall system. The force generated by the virtual environment is calculated from Eqs. (4.7) and (4.8):

$$F_d^n = Z_E^n v_{ee}^n \qquad (4.15)$$

where

$$Z_E^n = \frac{Z_e^n Z_c^n}{Z_e^n + Z_c^n} \qquad (4.16)$$

Replacing the desired force in Eq. (4.10) with the above formula and implementing a controller for compensation of device dynamic as in Eq. (4.11) result in the overall dynamic equation including VE:

$$\begin{bmatrix} F_h \\ F_{ee} \end{bmatrix} = \begin{bmatrix} Z_h + Z_d & H_f(Z_E^n - Z_d^n) \\ Z_d & H_f(Z_E^n - Z_d^n) \end{bmatrix} \begin{bmatrix} v_{ee} \\ v_{ee}^n \end{bmatrix} \qquad (4.17)$$

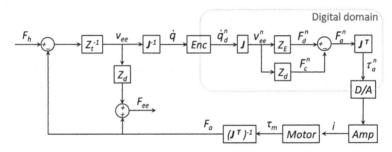

Fig. 4.4 Block diagram of haptic rendering system as a whole

The corresponding block diagram of the whole dynamic system is shown in Fig. 4.4. Using Eq. (4.17) and Eq. (4.4), a general equation for haptic interaction can be obtained:

$$F_{ee} = \left(Z_d - H_f Z_d^n H_v + H_f Z_E^n H_v\right) v_{ee} \qquad (4.18)$$

4.3 Categorization of Physical Evaluation

The physical evaluation methodology is categorized according to the three distinct properties of a robotic system: unpowered, powered and controlled system properties (see Fig. 4.5). Unpowered system properties can be identified without actuating the system. They are affected neither by the electronics nor by the control structure. Pure mechanism and structure design determines them. On the other hand, powered system properties define the capabilities of device actuation and sensing. The actuation and sensing flows are illustrated in Fig. 4.5. Investigating the input and output flow separately (without any loop) reveals the quality and capacity of these channels. The properties of the powered system is investigated without any control algorithm implemented. Finally, we can specify the whole controlled system performance by considering the loop formed by the human, mechanical structure, sensing and actuation.

Considering that the mechanical interaction takes place between the user and the end effector, measurements related to the overall system should be taken at this place.

4.4 Unpowered System Properties

Unpowered system properties can be identified without actuating the system. They are affected neither by electronics nor by control structure. Pure mechanism and structure design determines them. Figure 4.6 shows the input-output relationship of the unpowered system.

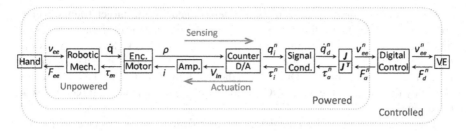

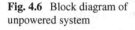

Fig. 4.5 Detailed flow chart representation of haptic rendering and the categorization of the physical evaluation methodology

Fig. 4.6 Block diagram of unpowered system

4.4.1 Kinematics

Like any other robotic mechanism, a force-feedback device is also designed for motion. Therefore, kinematics which covers describing the pose (q, i.e. position and orientation) and velocity of the mechanism is the basic tool to define the motion capabilities and boundaries. Besides, it is the base of the robot design for further actuation and sensor selection and control.

In order to describe the motion of a mechanism, both the joint space and task space analysis are crucial. The relationship between joint space velocities and task space velocities are represented by a matrix called *Jacobian* (J):

$$v = J(q)\dot{q} \tag{4.19}$$

where v is the velocity vector of the end effector and $\dot{q}$ is the joint rates vector. The Jacobian is the function of the pose and the basic component of forward and inverse kinematic analysis.

Kinematic performance of a robotic structure can be analyzed in terms of degrees of freedom, mechanical singularities, kinematic isotropy, workspace, manipulability and dexterity indices.

4.4.1.1 Workspace

Probably the most important property of any mechanical manipulator is its workspace since the workspace requirements are given by the user's task specification. Since the degrees of freedom (DOF), configuration of the kinematic chain and the workspace shape and volume play an important role in the task execution, the manipulator's freedom of motion defining its workspace should allow users to achieve the required tasks. Besides, there should not be any mechanical singularity within the workspace.

Degrees-of-freedom specifications are determined by the target frames which are set to be reached by the manipulator's end effector. However, since a haptic interface is not only a manipulator but also a force-feedback device, degrees-of-freedom specifications for haptic interfaces are given in terms of *passive* and *active* DOF. The number of passive DOF shows the freedom of motion of the end effector driven by the user and the active DOF is the number of independent force feedback directions.

The configuration of the kinematic structure defines the workspace shape and volume. Three different types of structure are considered: serial, parallel and hybrid.

The workspace representations are categorized as the *reachable* and the *dexterous* workspace [2]. The reachable workspace is defined to be the set points that can be reached by the end effector. However, the dexterous workspace consist of the points reached by arbitrary orientations. Depending on the kinematic configuration, the dexterous workspace may not exist or may be concurrent. In workspace analysis it is customary to represent the workspace graphically as an envelope or a basic geometrical shape.

4.4.1.2 Dexterity of a Manipulator

Dexterity is the ability of a manipulator to move and apply forces and torques in arbitrary directions [2]. Several performance metrics are proposed to characterize the dexterity. In this context, Yoshikawa [38] introduced the term *manipulability* which quantifies the manipulator's ease of arbitrarily changing position and orientation for a given posture. The manipulability is defined as

$$\mu = \sqrt{\det\left(J J^T\right)} \tag{4.20}$$

Salisbury and Craig [29] proposed a measure of distortion due to the Jacobian. The *condition number* of J is the ratio of the largest to the smallest singular values of J

$$\kappa_2 = \frac{\sigma_M}{\sigma_m} \tag{4.21}$$

which can also be calculated as

$$\kappa_2 = \|J\| \|J^T\| \tag{4.22}$$

The condition number approaches infinity near manipulator singularities. Thus, it is more convenient to use the reciprocal of the condition number, $1/\kappa$ as it is bounded between 0 and 1. As the condition number depends on the pose, it gives a local information about the dexterity. In order to obtain a more general metric, the *global conditioning index* (GCI) is introduced which is the integration of the condition number over the whole workspace [16]:

$$\eta = \frac{\int \frac{1}{\kappa} dW}{\int dW} \tag{4.23}$$

The GCI represents the movement isotropy over the workspace. The closer the GCI to unity value, the more even "feel" through the workspace [37].

4.4.2 Elastostatics

Elastostatics is the study of elasticity under static equilibrium conditions. Under this category, the stiffness of a robotic structure (i.e. *structural stiffness*) is examined by measuring the deformation of the structure under applied loads in static equilibrium.

4.4.2.1 Structural Stiffness

In robotic systems, high structural stiffness are desired because stiff transmission including links and joints provides reliable calculation of the end effector position and better structural response under dynamic loading. For haptic interfaces, structural stiffness measurements can be made at the endpoint when the joints are blocked. The deflection of the overall system at the endpoint are measured by a dial indicator while the system is loaded with a dynamometer at the same point.

4.4.3 Dynamics

4.4.3.1 Structural Dynamics

Characterizing the dynamic behavior of a robot is important to find structural resonant modes and frequencies. Experimental modal analysis provides useful tools to extract the structural dynamics such as vibration modes, resonant frequencies and damping characteristics. Typically, an accelerometer is attached to the structure to measure the resulting motion due to the input excitation load. Then, the frequency content of the signals is analyzed by the Fast Fourier Transform (FFT) algorithms.

The following excitation methods are generally used to extract the structural dynamics:

1. Impact testing: This test is often preferred because of simple equipment (e.g., hammer) usage and fast procedure. A load cell equipped hammer and an accelerometer on the structure are enough for full characterization of structural dynamic response. In case a simple hammer without any sensor is used, it is still possible to extract the resonant mode frequencies.
2. Shaker testing: Using a shaker attached to the structure, periodic excitations are provided and the response is measured by one or multiple accelerometers. Commonly used excitation signals are sine wave, sine sweep (or chirp) and white noise (random or pseudo random). The main advantage of the sine wave excitation signal is to put all the energy into the structure at a specific frequency that results in accurate measurements. However, since a certain amount of frequencies must be visited to obtain a meaningful response, this method is not time efficient. On the other hand, the sine sweep covers a wide range of frequencies in the spectrum in a relatively short period. Nevertheless, the most efficient method for shaker testing is to use a random signal which excites all the resonances of the structure in a random manner.

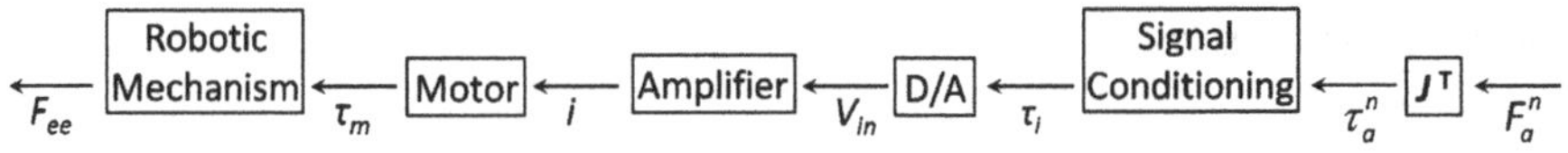

Fig. 4.7 Block diagram of actuation system. For the complete haptic system flow, see Fig. 4.5

4.4.3.2 Generalized Inertia Ellipsoid (GIE)

The *generalized inertia ellipsoid* (GIE) concept is introduced by Asada [3] as the geometric shape defined by

$$G = J^{-T} M J^{-1} \tag{4.24}$$

which characterizes the dynamic capability of a manipulator. In terms of haptic interaction, this metric can be seen as the inertia that is felt by the human while moving the end effector.

4.4.3.3 Acceleration Radius

Graettinger and Krogh [17] introduced the concept of *acceleration radius* as the maximum guaranteed magnitude of acceleration that can be achieved by the end effector over the entire operating region. For any manipulator, a maneuverability set can be defined which is the set of end-effector accelerations and bounded by the actuators' torque limits. The acceleration radius is the radius of the largest sphere in this maneuverability. Therefore, it is a measure of the size of this maneuverability set that can be used for comparison and design.

4.5 Powered System Properties

Powered system properties define the capabilities of device actuation and sensing. Investigating the input and output flow separately (without any loop) reveals the quality and performance of these channels. The properties of the powered system do not depend on any closed-loop control.

4.5.1 Actuation Capabilities

The actuation system includes all the electro-mechanical elements of the haptic interface from the virtual environment to the human user. These elements and the signal flow direction are shown in Fig. 4.7. Each element can affect the transfer function between the desired force signal and the output force and motion at the end effector.

4.5.1.1 Static Response

One of the most common performance specifications provided by manufacturers is the maximum output capability of a device in static conditions. In the case of haptic interfaces, the most relevant is of course maximum output force. There are two general ways to specify this: transient *peak force* and *maximum continuous force*. Since the first one is limited in time due to heat dissipation and actually short in its duration, the maximum continuous force is a better benchmark metric for haptic devices.

In order to measure the continuous output force at the end effector, a force sensor is attached to the end effector. A slowly increasing and decreasing ramp input is commanded to the actuators. Thus, an output versus input force graph is obtained. This input-output curve is also called calibration curve of a device and it shows the degree of *linearity* of the actuation [9]. It is possible to extract output force limits from this graph such as *maximum continuous force* due to saturation, *dead zone* due to stiction and Coulomb friction and *hysteresis* due to backlash, loose components and friction. The slope of the linear part of the calibration curve is called *sensitivity*, that is the change in output for a unit change in input. High sensitivity is desirable for any actuator. The limit of the dead zone gives the *minimum force* that can be generated by a haptic device. This is also known as *static friction breakaway force*, *stiction* or *backdrive friction*. This value should not be confused with device resolution. Resolution of an actuation system is the smallest change in an input signal that can be accurately reproduced at the output in the linear operating range. So one can say that *output force resolution* of a haptic interface is the smallest incremental force that can be generated in addition to the minimum force. The force resolution can be measured while a ramp force with a bias to overcome the dead zone is being given to the actuators. The smallest change is recorded as the force resolution. One of the important processes that affects force resolution is the digital to analog (D/A) conversion. Therefore, the *D/A resolution* should be stated as well since the whole actuation chain is measured.

The input-output curve also gives us a range of operation with a required linearity and accuracy. The *dynamic range* of a haptic device is expressed as the ratio of the maximum continuous force to the minimum force. It is often given in decibels (dB). The force dynamic range defines the extent of controllable force which has a lower limit at the friction level and an upper limit at the motor saturation. This is also known as *force depth*, i.e., maximum force divided by backdrive friction [34].

4.5.1.2 Frequency Response

A robotic system can be described by a model or a transfer function after an experimental identification is applied to input and output signals. Frequency domain transfer function models, in particular, are of interest for representing and analyzing actuators, sensors, structure and controllers, in other words, all components of a robotic system. Investigating each component is important because the performance of a system depends on the component error. Frequency analysis plays an important role in identifying the component behavior. The identification procedure relies on the FFT analysis resulting in the transfer function between the excitation and the response signals with respect to frequency which is generally represented by a Bode diagram. This information is then used to develop control architecture.

Operating bandwidth (or simply, bandwidth (ω_b)) provides the useful frequency range within which a system can operate. It is also called *instrument bandwidth*. It defines the maximum speed or frequency at which the instrument is capable of operating [9]. For a robotic system, the bandwidth corresponds to the speed of response at a given excitation: the higher the bandwidth, the faster the response. For an impedance type haptic interface, the operating bandwidth of the actuation system is determined through the commanded force. Therefore, we can also call this *force bandwidth*.

The bandwidth is calculated from the transfer function in frequency domain. There are two main definitions of the bandwidth that are related to the Bode diagram:

- the frequency at which the gain plot is not flat anymore (flat region is known as *useful frequency range*)
- the frequency at which the gain drops −3 dB from the zero frequency level. This is called *half-power bandwidth*.

In order to obtain the frequency response of a haptic interface at the end effector, a pseudo-random or sine sweep force input to the actuator is sufficient (for more detailed description of the excitation signals see Sect. 4.4.3.1). However, boundary conditions at the end effector and signal to be measured differ based on the transfer function of interest. The measurement conditions play an important role in determining the dynamic response as discussed in [18]. We can categorize the boundary or interaction conditions as follows (see also Fig. 4.8 for the illustration of the experimental conditions):

1. Open end: The detached end effector is free to move, therefore, no force measurement is possible at the tip. An accelerometer attached to the end effector measures output acceleration. This experiment will result in similar behavior as of the structural dynamics experiment explained in Sect. 4.4.3.1. However this time, it is possible to analyze the response which is affected not only by the mechanical structure dynamics but also by the electronics and signal conversion. Analyzing the behavior between the output velocity and the input force gives the modified admittance Bode diagram (as explained in Sect. 4.2 with Eq. (4.12)). This can be used to develop a force control strategy. Still, the open end boundary condition does not fully represent the normal operating conditions which requires human to be in the loop.
2. Constrained
   a. Fixed end: The end effector is firmly clamped to a rigid, stationary constraint and movement is not allowed. Force output is measured by a sensor between the constraint and the end effector. This measurement will provide the force transfer function (see Eq. (4.13) in Sect. 4.2) under static conditions and hide

Fig. 4.8 Boundary or interaction conditions for the frequency response measurements. A 1-DOF haptic interface is illustrated

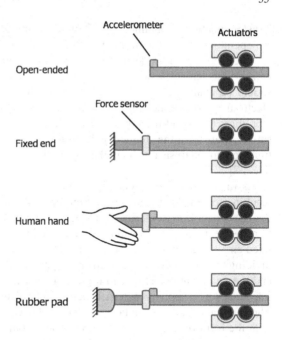

noise generated due to the dynamic behavior. Therefore, it is not really correct to rely on the frequency response of a haptic interface under static conditions which is never achieved during device use by a human operator.

b. Human hand: A human user grasps the end effector and holds it in the operating position (i.e., middle of the device workspace). Since the device is coupled with the human hand and arm, we can conclude that the dynamic response of the system will be the closest to the normal operating conditions. This measurement may also be called human-in-the-loop experiment. A derivation of the force transfer function under dynamic conditions is possible by measuring the force output at the end effector. In addition, measuring the velocity of the end effector will result in coupled human hand and device impedance of the system. The problem of using human in the experiments is the nonrepeatability of the conditions.

c. Rubber pad: A polymer sample resembling the tissue and/or grasp behavior of the human hand is placed between the end effector and the rigid contact. Hayward and Astley [18] proposed this solution in order to obtain meaningful boundaries representing the actual operating conditions. However, the problem is to find the right silicon or rubber sample which has similar impedance characteristics with the human hand. Force or acceleration output can be measured at the end effector.

In order to find the transfer function from the load on the actuator (F_a) to the end effector movement (v_{ee}) or force (F_{ee}), the excitations signals can be applied to the actuators as input currents (see Fig. 4.7). After the FFT analysis, the nonparametric transfer function is obtained as a Bode diagram. This also enables us to calculate the operating bandwidth of the interface.

Since the flow from desired input to realized output is degraded by the components of the actuation system, each component and corresponding transfer function should be analyzed separately. First of all, the frequency analysis of the driving electronics gives us the *amplifier bandwidth*. A sweep voltage is commanded to the amplifiers (V_{in}) and the output voltage across the current sensing resistor is measured. Apart from the frequency domain Bode diagram for bandwidth calculation, time domain analysis results in the signal distortion due to the amplifier nonlinearities [28].

As impedance is the dynamic relation between a given velocity input and the resulting force provided by a haptic interface, it comprises all the resistance against the movement due to inertia, damping, Coulomb friction etc. There are two ways to measure the *output impedance* of a haptic interface. The first one is to perform a fixed end and an open end experiment. The output impedance can be calculated by dividing the force output of a fixed end experiment by the velocity output of an open end measurement (as explained in Sect. 4.2). The other way to measure the output impedance is to apply an external excitation or disturbance by the end effector side. The external excitation can be applied by a shaker or a human user. The disturbance is in the form of a tapping or jiggling when a human user exerts it. It is generally band limited. Although voluntary human hand motion is limited to 4 Hz, periodic movements and tapping can produce higher frequencies up to 8 Hz (see Table 3.4). An example of this phenomenon is the drum roll and explained by the limit cycle [19, 22].

Although the bandwidth is a good estimator of the speed of a system, it does not capture the degree of signal distortion due to losses such as friction and backlash. It is better to compare the desired input and actual output signal in the time domain. On top of that, a measure called *force fidelity* has been introduced by Morrell and Salisbury [26]. It captures quantitatively the degrees of the force distortion from a true sinusoidal signal. Based on RMS error between the sinusoidal input signal and measured output, the force fidelity is defined as [35]

$$\text{force fidelity} = \left(1 - \frac{\text{var}(y - R)}{\text{var}(y)}\right) \qquad (4.25)$$

where y is the scaled and sampled true sine and R is the sampled output sine. The value of 1.0 represents an undistorted signal. The force fidelity is calculated for the frequencies of interest and plotted against frequency. It is also possible to simultaneously display the bandwidth and force fidelity data on a graph called force performance space which shows the amplitudes and frequencies at which the RMS error is less than 10 % [26].

4.5.1.3 Step Response

Apart from the frequency response, specifications in the time domain also give useful information about a robotic system's performance. In general, a step input is used to quantify the time domain performance specifications, such as rise time, percentage overshoot and settling time. The *rise time* (T_r) is the time at which the system passes the steady-state value of the response for the first time. It is related to the speed of response and the bandwidth of the system. In general, the following relation is used to determine the bandwidth:

$$\omega_b = \frac{0.35}{T_r} \tag{4.26}$$

Settling time or percentage overshoot represents the damping present in the system and the degree of stability.

Since the aim of this experiment is to quantify the relation between force step input and force output at the end effector (i.e., open-loop force control behavior), normal operating conditions of a haptic interface is desirable. Therefore, either a human user grasps the end effector or a rubber pad is used to constrain the movement of the tip.

Actuation accuracy and precision are also important performance metrics in the time domain for haptic interfaces. *Force output accuracy* indicates the degree of closeness of the force output to the desired value, which can also be called steady-state error. *Force precision*, on the other hand, indicates how repeatable the steady-state values are. Precision is also called repeatability and often indicated as ± or standard deviation.

The accuracy and repeatability should be measured under normal operating conditions and within the dynamics range of the device. Since it is difficult to obtain a repeatable measurement condition when human grasps the end effector, the best is do the experiments either with the rubber pad or rigid constraint. A constant value force is commanded and several force measurements are performed. Deviation from the desired value is plotted against number of measurement.

4.5.1.4 Impulse Response

Similar to the step response, the impulse response of a system gives information about not only damping and stability but also acceleration capability. *Peak speed* and *maximum acceleration* specifications are especially important for haptic interfaces because rendering unilateral contacts highly depends on the speed of the system. Hence, high acceleration is desirable. In order to measure the peak speed and maximum acceleration, an approximate impulse (e.g., a square pulse of 10 ms with a magnitude of peak force [18]) is applied and the end effector acceleration is measured by an accelerometer while the system is open-ended. Since the acceleration capability depends on the posture of the device, the measurements should be performed when the device is in the normal operating position.

A dimensionless measure, the *structural deformation ratio* (SDR), is proposed by Moreyra and Hannaford [25] in order to quantify the structural distortion during an impulse test. SDR is the ratio of the measured joint velocity to the theoretical velocity due to an impulse excitation and it is a measure of the flexibility of a device.

Fig. 4.9 Block diagram of sensing system. For the complete haptic system flow, see Fig. 4.5

4.5.2 Sensing Capabilities

In this section, the sensing capabilities of a haptic interface are investigated. Therefore, the haptic interface is considered as an input device or a measurement instrument. The input flow through this instrument, from a human user to the virtual environment, is shown in Fig. 4.9. Since impedance type haptic devices are investigated, the measurand is the motion of the end effector manipulated by a human user. The properties of each of these components define the overall sensing capabilities. De Silva [9] sorts the characteristics of a perfect measurement device as fast response, high gain and sensitivity, drift-free (i.e., stable) output, linearity and high input impedance.

4.5.2.1 Static Response

The *calibration curve* shows the relation between a measurand and the output of a sensor. In order to obtain the position calibration curve for a haptic interface, the end effector is positioned slowly to various known dimensions by a highly accurate and precise positioning instrument and the output of the sensor is recorded. The slope of the calibration curve gives the *sensitivity* of position sensing and a high value is desirable. Changing the direction of the positioning movement, *hysteresis* in the sensing due to backlash can be determined. In addition, the smallest change of position which can be detected by sensors is called *position resolution*. It is highly related to the counter resolution as well. Higher sensor resolution is desirable to eliminate quantization problems. It is also important especially if position measures is differentiated to provide velocity or acceleration estimates for virtual damping and mass [8]. Unless analog sensors dedicated to velocity and acceleration measurement (such as tachometers, accelerometers) are used, higher velocity or acceleration resolution can be achieved either by maximizing position resolution or using an appropriate filter.

The *dynamic range* of position sensing can be quantified using the position resolution and the workspace dimensions. The workspace provides the upper limit for the dynamic range of sensing.

The static measurements can be repeated several times to obtain *position measurement accuracy* and *precision* as well. Repeatability of the measurements determines the precision of position readings. Instrument accuracy is related to the worst accuracy obtainable within the dynamics range of the instrument in a specific operating environment [9].

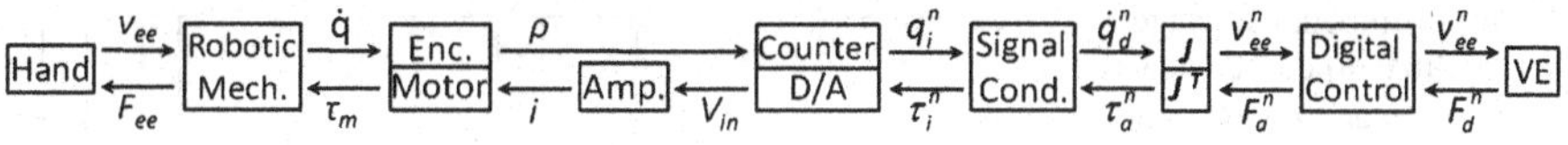

Fig. 4.10 Block diagram of controlled system as a whole

4.5.2.2 Frequency Response

Similar to the dynamic analysis of actuation described in Sect. 4.5.1, the frequency response of the sensing provides useful information about the frequencies for reliable measurements, hence, reliable operating conditions. If a haptic interface is considered also as a sensing instrument, the operating bandwidth is also limited to the useful frequency range of accurate measurements or in other words the *sensor bandwidth*. Therefore, it is desirable to obtain the frequency response of position sensing when a periodic position is commanded to the end effector. In addition, signal distortion in the sensed signal due to nonlinearities can be quantified using Eq. (4.25).

4.6 Controlled System Properties

The whole controlled system performance is evaluated by considering the loop formed by the human user, mechanical structure, sensing and actuation and virtual environment (see Fig. 4.10).

4.6.1 Impedance Range

An impedance type haptic interface generates forces based on a velocity input given at the end effector. Therefore, the range of impedances that can be rendered by a haptic interface should be as wide as possible. Considering the impedance distribution over frequency, this range is also called *Z-width* of a haptic display which is the dynamic range of impedances that can be passively rendered [8, 36]. The Z-width is illustrated in Fig. 4.11. This range is bounded by a minimum impedance at the lower limit. The *minimum impedance* represents the relation between the least resisting force for a given velocity input when a free space is simulated and system is compensated for dynamic disturbances such as inertia, damping, friction etc (i.e., desired output force is zero):

$$\frac{F_{ee}}{v_{ee}} = Z_d - H_f Z_d^n H_v \tag{4.27}$$

Therefore, it's also known as free impedance. It comprises all the resistance due to remaining uncontrolled disturbances after system is controlled for reducing the output impedance. The upper limit of the impedance range, *maximum impedance*,

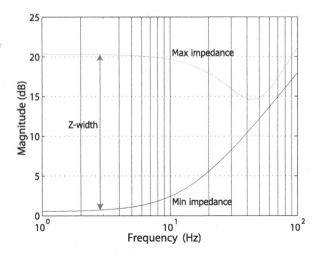

Fig. 4.11 Impedance range or Z-width defines a device's ability to render a wide range of haptic stimuli

is the highest possible gain that can be rendered without any human induced control instabilities by voluntary motion.

$$\frac{F_{ee}}{v_{ee}} = Z_d + H_f Z_c^n H_v \qquad (4.28)$$

The *haptic transparency* implies the fact that the user should not perceive the haptic interface at all [24]. Therefore, minimum impedance gives a quantitative measure of haptic transparency. Another way to measure transparency is to compare desired or simulated impedances with the actual or realized ones.

Two ways to measure the output impedance of a haptic interface are explained in Sect. 4.5.1. These steps can be well applied to measure the achievable minimum impedances under active control. Generally preferred method is that a human user jiggles the end effector while the interface renders a zero impedance or zero output force. As the human hand motion is limited to 10 Hz (see Table 3.4), the resulting minimum impedance will be band limited.

The maximum stable impedance is usually measured using a virtual wall simulation which has been demonstrated to be a reliable benchmark test for haptic interfaces. Unilateral constraints provide a boundary where the interface should render a sharp transition from minimum to maximum impedance. The actual impedance of the stiff virtual wall is determined by tapping the wall and measuring the contact force and velocity. This excitation can be considered as an impact. Analyzing the resulting force and velocities using a system identification method provides the maximum impedance with respect to frequency.

4.6.2 Control Bandwidth

The operating bandwidth of a haptic interface is bounded by control architecture speed. The control bandwidth provides the frequency range where a haptic interface can be controlled to operate accurately. For an impedance type haptic interface, we can specify an *impedance control bandwidth* which is determined when the system is under control. The desired impedance is set to the maximum impedance of the device and the virtual trajectory is given as a sweep sine. The transfer function from the desired force (i.e., desired position minus actual position times max impedance) to the output force gives the frequency response of a haptic interface under impedance control. Output force measurements can be made while either a human user grasps the end effector or a rubber pad is used to constrain the movement of the tip.

4.7 Physical Performance Metrics

A summary of the physical performance metrics is given in Table 4.2. Considering the exhaustive physical experiments, the number of parameters that need to be quantified may be reduced considering the target application of a device. This issue is further discussed in Sect. 4.8 and Chap. 8.

4.8 Conclusions

The goal of this chapter was to investigate physical performance measures and define their experimental methods. Variety of the measures, their dependency on boundary conditions and testing methods were discussed.

Considering the numerous physical experiments, to run all the tests for full device characterization is exhaustive. However, considering the basic actuation and sensing flows and the loops formed as shown in Fig. 4.5, number of the most relevant physical evaluation methods can be reduced to three: static response measurements for actuation and sensing and the impedance range (i.e., Z-width) measurements. First two will give the input-output curves which are crucial to define minimum and maximum limits of a device. The latter will define the device's ability to render a wide range of haptic stimuli. Since the impedance range measurement takes into account all components of haptic interaction (i.e., the virtual reality environment, the haptic interface and the user behavior), it is a very comprehensive metric including dynamic behavior.

The biggest challenge of the device characterization is not defining these measures. The challenge is in the practice. First of all, lack of proper equipment limits the applicability of the measurements. Second, if the system being measured is highly nonlinear, then it is difficult to obtain meaningful results. Third, most of the measurements are position dependent. Therefore, providing performance measures only on the normal operating condition (i.e., in the middle of the workspace) might not necessarily mean that the device will perform the same every point in the workspace.

Table 4.2 Physical performance metrics

System	Type		Metric
Unpowered	Kinematics	Workspace	Reachable
			Dexterous
		DOF	Passive
			Active
		Structure	Serial
			Parallel
			Hybrid
		Dexterity	Manipulability
			Condition number
			Global conditioning index
	Elastostatics		Structural stiffness
	Dynamics		Structural dynamics
			Generalized inertia ellipsoid
			Acceleration radius
Powered	Actuation	Static	Peak force
			Continuous force
			Minimum force
			Hysteresis
			Sensitivity
			Output force resolution
			D/A resolution
			Dynamic range
		Frequency Response	Force bandwidth
			Useful frequency range
			Amplifier bandwidth
			Output impedance
			Force fidelity
		Step Response	Rise time
			Settling time
			Overshoot
			Output force accuracy
			Force precision
		Impulse Response	Peak speed
			Peak acceleration
			Structural deformation ratio
	Sensing	Static	Sensitivity
			Hysteresis

Table 4.2 (continued)

System	Type		Metric
			Position resolution
			Dynamic range
			Position measurement accuracy
			Precision
		Frequency Response	Sensor bandwidth
Controlled	Impedance Range		Min impedance
			Max impedance
			Z-width
	Control Bandwidth		Impedance control bandwidth

References

1. Adams, R.J., Hannaford, B.: Stable haptic interaction with virtual environments. IEEE Trans. Robot. Autom. **15**(3), 465–474 (1999)
2. Angeles, J., Park, F.C.: Performance evaluation and design criteria. In: Siciliano, B., Khatib, O. (eds.) Springer Handbook of Robotics, pp. 229–244. Springer, Berlin (2008)
3. Asada, H.: A geometrical representation of manipulator dynamics and its application to arm design. J. Dyn. Syst. Meas. Control **105**(3), 131–142 (1983)
4. Bergamasco, M., Frisoli, A., Avizzano, C.: Exoskeletons as man-machine interface systems for teleoperation and interaction in virtual environments. In: Ferre, M., Buss, M., Aracil, R., Melchiorri, C., Balaguer, C. (eds.) Advances in Telerobotics. Springer Tracts in Advanced Robotics, vol. 31, pp. 61–76. Springer, Berlin (2007)
5. Brooks, T.L.: Telerobotic response requirements. In: IEEE International Conference on Systems, Man and Cybernetics, pp. 113–120 (1990)
6. Cavusoglu, M.C., Feygin, D., Tendick, F.: A critical study of the mechanical and electrical properties of the phantom haptic interface and improvements for high-performance control. Presence Teleoperators Virtual Environ. **11**(6), 555–568 (2002)
7. Chapuis, D.: Application of ultrasonic motors to MR-compatible haptic interfaces. PhD thesis, EPFL, No. 4317 (2009)
8. Colgate, J.E., Brown, J.M.: Factors affecting the Z-width of a haptic display. In: IEEE Int. Conf. Robotics and Automation, pp. 3205–3210 (1994)
9. De Silva, C.W.: Sensors and Actuators and Control System Instrumentation. Taylor & Francis, CRC Press, Boca Raton (2007)
10. Ellis, R., Ismaeil, O., Lipsett, M.: Design and evaluation of a high-performance haptic interface. Robotica **14**, 321–327 (1996)
11. Eppinger, S.D.: Modeling robot dynamic performance for endpoint force control. PhD thesis, MIT (1988)
12. Faulring, E.L., Colgate, J.E., Peshkin, M.A.: The cobotic hand controller: design, control and performance of a novel haptic display. Int. J. Robot. Res. **25**, 1099–1119 (2006)
13. Frisoli, A., Bergamasco, M.: Experimental identification and evaluation of performance of a 2 dof haptic display. In: Proc. of IEEE International Conference on Robotics and Automation, vol. 3, pp. 3260–3265 (2003)

14. Gassert, R., Moser, R., Burdet, E., Bleuler, H.: MRI/fMRI-compatible robotic system with force feedback for interaction with human motion. IEEE/ASME Trans. Mechatron. **11**(2), 216–224 (2006)
15. Gillespie, R.B.: Haptic interface to virtual environments. In: Kurfess, T. (ed.) Robotics and Automation Handbook. CRC Press, Boca Raton (2005)
16. Gosselin, C., Angeles, J.: A global performance index for the kinematic optimization of robotic manipulators. J. Mech. Des. **113**(3), 220–226 (1991)
17. Graettinger, T.J., Krogh, B.H.: The acceleration radius: a global performance measure for robotic manipulators. IEEE J. Robot. Autom. **4**(1), 60–69 (1988)
18. Hayward, V., Astley, O.: Performance measures for haptic interfaces. In: Robotics Research: The 7th International Symposium, pp. 195–207 (1996)
19. Hayward, V., Maclean, K.E.: Do it yourself haptics: part i. IEEE Robot. Autom. Mag. **14**(4), 88–104 (2007)
20. Hogan, N., Buerger, S.P.: Impedance and interaction control. In: Kurfess, T. (ed.) Robotics and Automation Handbook. CRC Press, Boca Raton (2005)
21. Hollerbach, J.M., Hunter, I.W., Ballantyne, J.: A comparative analysis of actuator technologies for robotics. In: The Robotics Review, vol. 2, pp. 299–342. MIT Press, Cambridge (1992)
22. Lawrence, D.A., Pao, L.Y., Salada, M.A., Dougherty, A.M.: Quantitative experimental analysis of transparency and stability in haptic interfaces. In: Proc. of ASME Dynamic Systems and Control Division, DSC, vol. 58, pp. 441–449 (1996)
23. McAffee, D.A., Fiorini, P.: Hand controller design requirements and performance issues in telerobotics. In: Fifth International Conference on Advanced Robotics, ICAR, vol. 1, pp. 186–192 (1991)
24. Moix, T.: Mechatronic elements and haptic rendering for computer-assisted minimally invasive surgery training. PhD thesis, EPFL, No. 3306 (2005)
25. Moreyra, M., Hannaford, B.: A practical measure of dynamic response of haptic devices. In: Proc. of IEEE International Conference on Robotics and Automation, pp. 369–374 (1998)
26. Morrell, J.B., Salisbury, J.K.: Parallel-coupled micro-macro actuators. Int. J. Robot. Res. **17**, 773–791 (1998)
27. Peer, A., Buss, M.: A new admittance-type haptic interface for bimanual manipulations. IEEE/ASME Trans. Mechatron. **13** (2008)
28. Salisbury, C., Gillespie, R.B., Tan, H., Barbagli, F., Salisbury, J.K.: Effects of haptic device attributes on vibration detection thresholds. In: Proc. of World Haptics'09, pp. 115–120 (2009)
29. Salisbury, J.K., Craig, J.J.: Articulated hands: force control and kinematic issues. Int. J. Robot. Res. **1**, 4–17 (1982)
30. Samur, E., Flaction, L., Bleuler, H.: Design and evaluation of a novel haptic interface for endoscopic simulation. IEEE Trans. Haptics (2011). doi:10.1109/TOH.2011.70
31. Stocco, L.J.: Robot design optimization with haptic interface applications. PhD thesis, University of British Columbia (1999)
32. Ueberle, M., Buss, M.: Design, control, and evaluation of a new 6 dof haptic device. In: IEEE/RSJ International Conference on Intelligent Robots and Systems, vol. 3, pp. 2949–2954 (2002)
33. Ueberle, M.W.: Design, control, and evaluation of a family of kinesthetic haptic interfaces. PhD thesis, Technische Universität München (2006)
34. Van der Linde, R.Q., Lammertse, P., Frederiksen, E., Ruiter, B.: The hapticmaster, a new high-performance haptic interface. In: Proc. of Eurohaptics'02 (2002)
35. Veneman, J.F., Ekkelenkamp, R., Kruidhof, R., van der Helm, F.C.T., van der Kooij, H.: A series elastic- and Bowden-cable-based actuation system for use as torque actuator in exoskeleton-type robots. Int. J. Robot. Res. **25**, 261–281 (2006)
36. Weir, D.W., Colgate, J.E., Peshkin, M.A.: Measuring and increasing z-width with active electrical damping. In: Proc. of IEEE International Symposium on Haptic Interfaces for Virtual Environment and Teleoperator Systems, pp. 169–175 (2008)
37. Yoon, J., Ryu, J.: Design, fabrication, and evaluation of a new haptic device using a parallel mechanism. IEEE/ASME Trans. Mechatron. **6**(3), 221–233 (2001)

38. Yoshikawa, T.: Manipulability of robotic mechanisms. Int. J. Robot. Res. **4**, 3–9 (1985)
39. Zinn, M., Khatib, O., Roth, B., Salisbury, J.K.: Large workspace haptic devices—a new actuation approach. In: Proceedings of the 2008 Symposium on Haptic Interfaces for Virtual Environment and Teleoperator Systems, HAPTICS'08, pp. 185–192. IEEE Computer Society, Washington (2008)

Chapter 5
Application to a Haptic Interface

Abstract In this chapter, the physical evaluation methods described in Chap. 4 are demonstrated on a two degrees-of-freedom haptic interface for surgical simulation. Since the whole physical evaluation methodology is very extensive and often requires special equipment, only some methods that are necessary to characterize the haptic interface are discussed here. After applying these methods, the physical evaluation of the haptic interface provided very useful information about the design and showed the points that need to be improved. This chapter is partially based on Samur et al. (IEEE Trans. Haptics, 2011, doi:10.1109/TOH.2011.70).

5.1 Apparatus: 2-DOF Haptic Interface for Colonoscopy Simulation

To address the need for higher fidelity and complexity in a colonoscopy simulator, we have designed a new haptic interface and instrumented a clinical colonoscope in order to integrate it with the software simulation for colonoscopy (MILX™ GastroSim) developed at CSIRO (see Fig. 5.1). Details of the novel design of the interface is given in [7] and more information about the software and the instrumented colonoscope can be found in [4–6].

5.1.1 Design Requirements

During a typical colonoscopy procedure, the colonoscope is inserted and rotated along the colon. Therefore, the linear and rotational workspace of a simulator should be practically unlimited for a realistic application. In addition, tool re-insertion is also required. Estimated forces and torques for a procedure performed with an adult size colonoscope are in the range of about ± 40 N and ± 1 Nm respectively [1].

E. Samur, *Performance Metrics for Haptic Interfaces*,
Springer Series on Touch and Haptic Systems,
DOI 10.1007/978-1-4471-4225-6_5, © Springer-Verlag London 2012

Fig. 5.1 2-DOF haptic
interface for colonoscopy
simulation. It provides force
feedback in translational and
rotational directions. The
embedded electronics on the
device provide a compact and
robust control solution

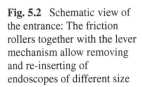

Fig. 5.2 Schematic view of
the entrance: The friction
rollers together with the lever
mechanism allow removing
and re-inserting of
endoscopes of different size

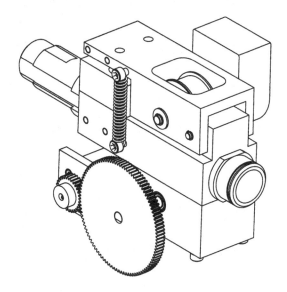

5.2 Physical Evaluation

Considering the above requirements, the haptic device was designed to work with
an original clinical colonoscope, ∅13 mm diameter Olympus CF-140L (Olympus
Corporation, Japan). The haptic interface tracks the position and orientation of the
colonoscope which is inserted in the simulator through a hole as shown in Fig. 5.2. It
also provides force feedback in translational and rotational directions by the colonoscope. The embedded electronics on the device, as shown in Fig. 5.1, provide a
compact and robust control solution. It is interfaced to a PC via USB, allowing
transmission at haptic rates with a plug-and-play solution.

5.2.1 Unpowered System Properties

The entrance mechanism of the haptic interface are shown in Figs. 5.2 and 5.3. The
haptic interface has 2 active degrees of freedom. It has an unlimited translational

Fig. 5.3 Schematic bottom view of the entrance: The V-shaped friction roller, which improves the contact between the friction rollers and the endoscope, is connected to a DC motor through a series of gears

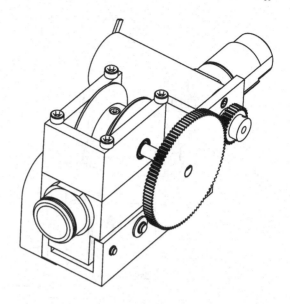

and rotational workspace. Force feedback is provided by a pair of friction rollers. The rollers are also used to track the axial displacement of the colonoscope. A lever structure which is constraint by a mechanical spring (see Fig. 5.2) ensures sufficient contact force on the endoscope while allowing smooth insertion and removal of different size of endoscopes.

The V-shaped friction roller as shown in Fig. 5.3 is used to balance the force acting on the endoscope for proper force feedback. It is connected to a DC motor through a series of gears. The other friction roller is engaged with an encoder for the tracking of the translational displacement. Decoupling of actuation and tracking systems avoids loss of position tracking data due to an unlikely slippage of the endoscope over the friction rollers. The rotational part is fixed on bearings and linked to a DC motor by means of a pair of gears.

5.2.2 Actuation & Sensing Capabilities

The actuation system consists of electrical motors and passive brakes to cover a large range of forces. RE25 DC motors (Maxon Motor, Switzerland) with a maximum output torque 28.8 mNm are used for active force feedback which is attached to the rotating part as shown in Fig. 5.3. An S90MPA-B15D19S magnetic particle brake (Sterling Instrument, USA), which has a maximum torque of 113 mNm, is chosen to apply resisting torques on the colonoscope. 500 CPT optical encoders are used to track linear and rotational displacements.

We have designed a custom mechanical brake as shown in Fig. 5.4. It is used in combination with the translational motor to render higher forces. A Maxon RE25 motor together with a Planetary Gearhead activates the mechanical brake consisting

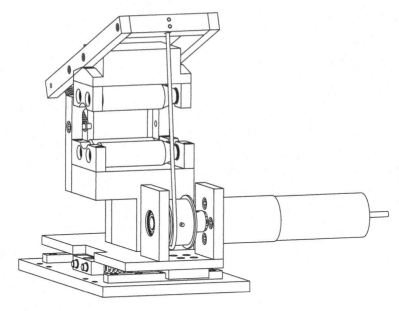

Fig. 5.4 Schematic view of the brake mechanism

of four rollers. When the brake is closed, all the cylinders compress the colonoscope, impeding the linear displacement, but still allowing rotational movement of the colonoscope.

The brake system is mounted on two linear guidance rails as shown in Fig. 5.4. Springs are placed on each side of the carriage in order to set the position of the brake system. An attempt of withdraw of the endoscope when the brake is activated results in a slight linear displacement which results in the release of the brake. This offers an alternative to force/pressure sensors to detect any change of force applied to the system by the operator.

5.2.2.1 Static Response

The experimental setup used for static response measurements is shown in Fig. 5.5. In this fixed-end boundary condition, a 120 N and 2.0 Nm capacity 6-DOF force/torque sensor (type: Mini40, calibration: SI-40-2, ATI Industrial Automation, Inc.) was used to measure output torque and force. In order to be able to attach the force sensor, the colonoscope was replaced by a 13 mm diameter and 350 g brass rod. The rod was also covered by a plastic heat shrink to obtain similar surface characteristics with the colonoscope. Data acquisition is performed at 1 kHz using two NI PCI-cards (one is dedicated to the force sensor) and the embedded control card of the haptic interface. The internal loop of the control card was set to 5 kHz to be able to generate high frequency excitations.

Fig. 5.5 The fixed-end boundary condition: a force/torque sensor is attached between the tip of the tool and the stationary base

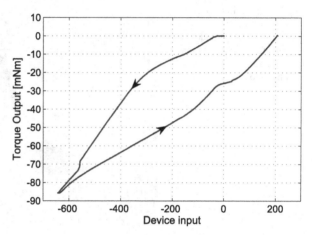

Fig. 5.6 Input-output curve of the rotation DOF. The maximum continuous torque produced by the DC motor is 85 mNm. The hysteresis loop shows high nonlinearity in this DOF

The input-output calibration curve of the device was obtained by commanding a ramp input to the motors. An example is shown in Fig. 5.6. High nonlinearity in the rotation DOF is obvious in this figure. The possible reasons of this nonlinearity are, first, the Coulomb friction introduced by the slip ring mechanism and second, backlash in the gears. The backlash can be seen in Fig. 5.7 as the sudden jump in the dead zone near zero. The calibration curves show that the device's maximum continuous force and torque are 5.4 N and 85 mNm, respectively. These maximum values are measured when the brakes are not turned on. The DC motors are set to this saturation level by the software in order to stay in the continuous operation range. When the brakes are activated, the maximum resistive force and torque go up to 80 N and 300 mNm, respectively.

The dead zone determines the minimum force (0.5 N) and torque (8 mNm). Using the maximum continues force and torque, the dynamic range is calculated as 20 dB when the brakes are not activated. With brakes, this value is as high as 40 dB. The sensitivity of the device is 0.01 N/*unit integer* for the translation and

Fig. 5.7 Input-output curve
of the rotation DOF showing
the dead zone. Due to the
stiction, the device does not
respond to the excitation
range ±40 (corresponding to
±8 mNm theoretical value of
the motor output).
Nonlinearity around zero is
due to the backlash of the
gears

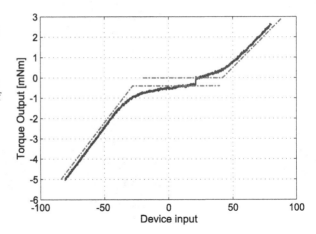

Fig. 5.8 Experimental setup
to obtain the position
resolution. A linear slide
micrometer is attached to the
tip of the force sensor and
then fixed to the base

0.2 mNm/*unit integer* for the rotation. The unit integer is a 12 bit integer representation in the controller card. By taking into account the linear ranges, the output force resolution of the device calculated as 9 mN and 0.12 mNm.

In order to measure the position resolution of the device experimentally, a new setup was installed. A linear slide micrometer was used to generate micron level movements (Fig. 5.8). The slide moved slowly until the encoders detected a change. Using this method, the position resolution was found as 0.045 mm which is consistent with the theoretical calculation.

5.2.2.2 Frequency Response

The frequency response of the actuation system for rotation DOF was measured by commanding a sweep input which has a frequency range from 1 Hz to 50 Hz. First, the fixed end boundary condition as shown in Fig. 5.5 and then the open end

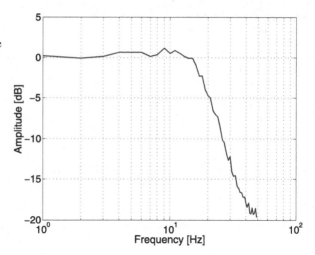

Fig. 5.9 Bode magnitude diagram of H_f for the rotation DOF. A sweep torque is commanded to the motors and the output torque is measured when the end effector is constrained

condition was imposed. As explained in Sect. 4.2, the fixed end boundary condition resulted in the transfer function in frequency domain from the input force to the force at the end effector (H_f, see Eq. (4.13)). The Bode diagram corresponding to the transfer function for rotation is shown in Fig. 5.9. It can be seen in this figure that the useful frequency range is up to 10 Hz and the force bandwidth is around 13 Hz.

The open-end experiment for the translation DOF was conducted using a $\pm 3g$ triple axis accelerometer (Type: ADXL335, SparkFun Electronics) attached to the tip of the tube. While a sweep input was given to the motor, accelerometer output was measured. For the rotation DOF, on the other hand, position readings from the encoder were used to calculate the velocity of the end effector. The transfer function from the input force to the velocity at the end effector results in the modified admittance of the device, Y_f, as given Eq. (4.12) in Sect. 4.2. The obtained Bode diagram for the rotation DOF is shown in Fig. 5.10.

The method proposed by Chapuis [2], which is explained in Sect. 4.2, was used to calculate the output impedance (Z_d) of the uncontrolled system. The force response of the fixed-end frequency experiment was divided by the velocity response of the open-end experiment using Eq. (4.14). This resulted in an impedance response bode diagram which is shown in Fig. 5.11. In order to develop a model-based force control strategy, the output impedance was then used to extract device dynamics. A second order mass-spring-damper model was fit to the experimental data. The obtained parameters of the model were 0.0011 kg m^2 for inertia, 6 Nm/rad for spring and 0.07 Nm s/rad for damping.

5.2.2.3 Impulse Response

The speed of the device was calculated by applying an approximate impulse and measuring the end effector acceleration while the system was open ended. The velocity and acceleration measures for a square wave of 50 ms with a magnitude of

Fig. 5.10 Bode magnitude diagram of Y_f for the rotation DOF. A sweep torque is commanded to the motor and the output velocity is measured when the end effector is free to move

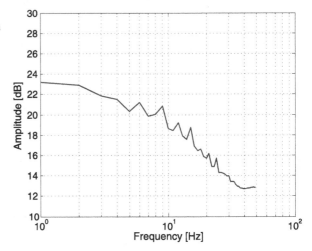

Fig. 5.11 Output impedance (Z_d) of the uncontrolled system in the rotational DOF. A second-order model is fit to the experimental data

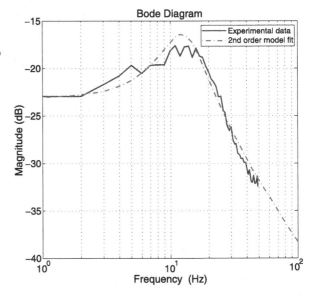

85 mNm (i.e., the maximum continuous torque) are shown in Fig. 5.12. Peak speed and maximum acceleration were 4 rad/s and 50 rad/s^2, respectively for rotation.

5.2.3 Controlled System Performance

Considering the 2nd order model parameters, we can conclude that the device has high output impedance as seen in Fig. 5.11. This shows that the uncontrolled systems suffers a lot in terms of transparency. The main reason for high impedance is

Fig. 5.12 Peak velocity and maximum acceleration for the rotation DOF

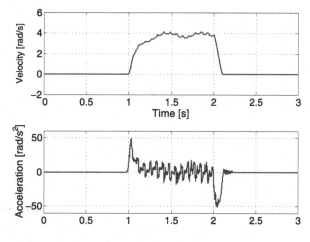

Fig. 5.13 Experimental setup to obtain the impedance range. While the user jiggles the end effector, velocity and force at the end effector are measured

the inertia of the tool and the friction in the device. Therefore, a model-based feed-forward control has been implemented using the damping parameter of the second-order model.

5.2.3.1 Impedance Range

The performance of the controller was evaluated by the impedance range (i.e., Z-width [3, 8]) measurements for the translation DOF. First, the minimum achievable impedances under the feed-forward control was measured with respect to frequency. To obtain the minimum impedance, a user jiggled the end effector while the VE was rendering a zero impedance (see Fig. 5.13). Since the human hand motion is limited to 7–8 Hz for movements such as periodic tracking and tapping (see Table 3.4), different frequency levels up to 10 Hz which can be achieved by voluntary human motion were targeted. Resulting velocities and forces at the end effec-

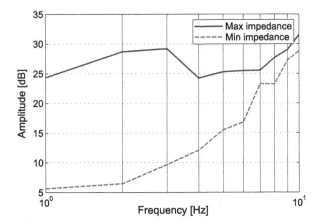

Fig. 5.14 Impedance range of the device for the translation DOF. The mechanical brake considerably increases the maximum stable impedance

tor were measured. Then, FFT analyses were performed on the measured data and the minimum impedance was calculated using Eq. (4.27). The resulting minimum impedance is shown in Fig. 5.14.

Compared to the uncontrolled system (please refer to Fig. 17 in [7]), the output impedance is considerable low. We can conclude that the device dynamics is well compensated with the model-based control, especially for the damping losses for low frequencies. As the inertia effect is not compensated by the control, the output impedance increases with frequency.

Since the friction and damping in the system are compensated at the low frequency region, we can assume that the device behaves as a pure mass in this region. Therefore, the minimum impedance of the device for low frequency region was interpreted in grams and calculated 112 gr.

As the second step to calculate the impedance range, maximum impedance measurements were performed for the translation DOF. In order to obtain the maximum impedances that can be rendered by the device without any instability, a virtual wall simulation was developed for the translation DOF. In order to obtain maximum stiffness, the mechanical brake together with the DC motor was activated. In this case, the stiffness value could go up to 2 N/mm. Actual impedance of the stiff virtual wall was determined by tapping the wall. Measured contact forces and velocities were analyzed using the FFT. Obtained maximum impedance with respect to frequency is given in Fig. 5.14. This graph also shows the impedance range of the device which is around 20 dB.

5.3 Conclusions

Overall, the physical evaluation of the colonoscopy device provided many useful information about the design and showed the points that should be improved. The experimental evaluation showed that the performance measures satisfy the design requirements summarized in Sect. 5.1 except the maximum torque requirement

(1 Nm). Limitation of 0.3 Nm maximum torque in the rotation DOF is due to the slip between the endoscope and the friction rollers.

It has been also shown that the device is subject to high nonlinearities such as backlash, Coulomb and viscous friction. The output impedance was high due to the friction in the uncontrolled system. Similarly, the force bandwidth of the system was relatively low. A model-based feedforward control has been implemented and the result showed that the control successfully compensates for the device dynamics. However, in order to reduce the backlash, the transmission system, which consists of gears, should be modified. One possibility could be to use belt drives or cable driven systems.

The inherent damping in the colonoscopy device helps to obtain higher stiffness values for the stable wall simulations, although it is a drawback for the transparency (see Eqs. (4.27) and (4.28)). This phenomenon was studied by Colgate and Brown [3] and they conclude that high damping in a system is desirable, as long as it is compensated for free movements.

References

1. Appleyard, M.N., Mosse, C.A., Mills, T.N., Bell, G.D., Castillo, F.D., Swain, C.P.: The measurement of forces exerted during colonoscopy. Gastroint. Endosc. **52**(2), 237–240 (2000)
2. Chapuis, D.: Application of ultrasonic motors to MR-compatible haptic interfaces. PhD thesis, EPFL, No. 4317 (2009)
3. Colgate, J.E., Brown, J.M.: Factors affecting the Z-width of a haptic display. In: IEEE Int. Conf. Robotics and Automation, pp. 3205–3210 (1994)
4. de Visser, H., Passenger, J., Conlan, D., Russ, C., Hellier, D., Cheng, M., Acosta, O., Ourselin, S., Salvado, O.: Developing a next generation colonoscopy simulator. Int. J. Image Graph. **10**(2), 203–217 (2010)
5. Hellier, D., Samur, E., Passenger, J., Spaelter, U., Frimmel, H., Appleyard, M., Bleuler, H., Ourselin, S.: A modular simulation framework for colonoscopy using a new haptic device. In: Proc. of the 16th Medicine Meets Virtual Reality Conference (MMVR) (2008)
6. Maillard, P., Flaction, L., Samur, E., Hellier, D., Passenger, J., Bleuler, H.: Instrumentation of a clinical colonoscope for surgical simulation. In: EMBS 2008, 30th Annual International Conference of the IEEE, pp. 70–73. Engineering in Medicine and Biology Society, Piscataway (2008)
7. Samur, E., Flaction, L., Bleuler, H.: Design and evaluation of a novel haptic interface for endoscopic simulation. IEEE Trans. Haptics (2011). doi:10.1109/TOH.2011.70
8. Weir, D.W., Colgate, J.E., Peshkin, M.A.: Measuring and increasing z-width with active electrical damping. In: Proc. of IEEE International Symposium on Haptic Interfaces for Virtual Environment and Teleoperator Systems, pp. 169–175 (2008)

Part III
Psychophysical Evaluation

Chapter 6
Performance Evaluation Based on Psychophysical Tests

Abstract A haptic interface is meant to be used by a human user. Its performance is highly affected by the user behavior and constraints. Application based or task specific evaluation methods have been proposed to test haptic interfaces during their proper use. This approach requires a human user to perform the task, thus it involves human dynamic and intention uncertainties. Although there is a variety of task specific evaluation approaches for haptic interfaces, it has not yet been possible to define a norm for meaningful device comparison and assessment. A proper evaluation procedure for haptic interfaces should link device performance measures to the limits of human perception in order to obtain device-specific limits. This chapter deals with psychophysical evaluation methodology. Testbed evaluation approach is applied to haptic interactions and a set of benchmark metrics are provided for haptic interfaces. We describe the methodology of seven testbeds and experimentally demonstrate usefulness of the testbeds on three commercial force-feedback devices. Finally, the general discussion at the end of this chapter summarizes the outcomes of the proposed psychophysical evaluation methodology.

6.1 Introduction

As haptic interfaces are intended to be used by a human user, human psychophysical limits provide minimum requirements for the haptic interaction. Since these limits are well studied and documented in the literature, human perception can also be used to evaluate the haptic interfaces. Hence, psychophysical evaluation approach is to apply human performance estimations to haptic interactions in order to have a systematic and complete evaluation method for haptic interfaces. We have reviewed and investigated a wide range of different psychophysical experiments: Peg-in-hole [9, 10, 29], tapping [5, 31], targeting [23], haptic training [1], joint tasks in a shared VE [2], hardness perception [17] and object recognition [6, 13, 14, 22, 24, 27, 30, 32] Based on these resources, we apply Bowman and Hodges' testbed evaluation approach [3] to haptic interactions and synthesize a set of evaluation testbeds. Careful repetition of described human factor studies leads to basic quantitative benchmark metrics for haptic interfaces [25].

E. Samur, *Performance Metrics for Haptic Interfaces*,
Springer Series on Touch and Haptic Systems,
DOI 10.1007/978-1-4471-4225-6_6, © Springer-Verlag London 2012

Fig. 6.1 Taxonomy of haptic modes

6.2 Testbed Evaluation Methodology

The main goal of the testbed evaluation approach [3] is to find generic performance characteristics. This yields general and complete results which can be applied to any VE application using the tasks studied within a testbed.

6.2.1 Categorization of Haptic Interaction Tasks

The first step towards a complete testbed is to gain an intuitive understanding of the generic interaction tasks and current techniques available for the tasks [3]. The initial evaluation experiences, that may be drawn from human factor studies in the literature, result in a taxonomy, independent outside factors/variables, and performance measures. Outside factors consist of task, environment, user and system characteristics which may individually affect user performance. Performance measures, on the other hand, include both quantifiable metrics and subjective performance values.

Kirkpatrick and Douglas [14] suggest a taxonomy of haptic modes. They divided the taxonomy into motor control and perception, which is shown in Fig. 6.1. This highest level distinction is also consistent with haptic interaction categories of human sensory-motor system. Jandura and Srinivasan [11] divided haptic interactions into exploration and manipulation which are dominated by sensory and motor system, respectively. The sensorimotor continuum introduced by Jones and Lederman [12] (see also Chap. 3) supports this taxonomy of the haptic modes.

Based on this initial categorization into motor control and perception, the next step is to define generic interaction tasks. Bowman and Hodges [3] propose that many interactions within a VE may be divided into three general tasks as *travel* (movement of one's viewpoint), *selection* (act of choosing a virtual object) and *manipulation* (task of setting the position and orientation of a selected object) (Fig. 6.2). They further brake down these tasks into subtasks and included the haptic feedback into their taxonomy. These generic interaction tasks correspond to the motor control mode of haptics and do not represent the perception mode directly.

As explained in Chap. 3, the three general methods in psychophysics, detection, discrimination, and identification, are used to determine the human thresholds of perception. They may also be interpreted as the generic interaction tasks of object exploration dominated by the perception mode of haptics (Fig. 6.3). Detection and discrimination involve the measurement of sensory thresholds of perception of a stimulus (e.g., position, velocity, force, pressure, stiffness, and viscosity) as shown in Fig. 6.3. In addition to the detection and discrimination, the third generic task in perception mode is identification, which involves categorizing stimuli. Since material properties (e.g., texture, hardness, and weight) are not the primary basis for

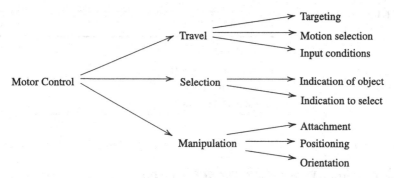

Fig. 6.2 Taxonomy of interaction tasks related to the motor control mode of haptics (the subtasks are adapted from Bowman & Hodges [3])

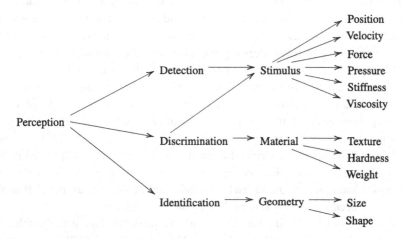

Fig. 6.3 Taxonomy of interaction tasks related to the perception mode of haptics. Detection and discrimination involve measurement of a stimulus

haptic object identification [15], identification of geometric dimensions and discrimination of material properties should be examined separately in order to properly evaluate the effectiveness of haptic interfaces on object recognition (Fig. 6.3).

After defining generic interaction tasks, corresponding interaction techniques can be determined. *Interaction techniques* combine natural human capabilities (particularly communication, motor, cognitive, and perceptual skills) with computer I/O devices [28]. The feedback techniques connect human sensory and motor system to haptic devices and include not only haptic but also visual and audio types. The haptic feedback can be divided into two interaction technique components as kinesthetic (or force) and tactile feedback since they are sensed by different receptors in the human body and, therefore, have different effects on the sensory-motor control. As a result, the main independent variable should be feedback technique including two components of haptic feedback in the testbed evaluation of haptic interfaces.

Table 6.1 Psychophysical performance metrics

Performance Metric	Unit
Index of Performance (*IP*)	bit/s
Information Transfer (*IT*)	bits
Absolute Threshold	
Differential Threshold (*JND*)	

6.2.2 Psychophysical Performance Metrics

Basic quantitative performance metrics for any task analysis are time and accuracy which are easy to measure. However, these two metrics may not be enough to quantitatively asses and compare the effectiveness of a haptic interface. Therefore, a simple measure characterizing the human-machine interaction is required. Fitts [7] applied the concept of information theory to the human motor control system and provided an *index of performance* (*IP*) which expresses the results of a movement as a performance rate in *bit/s*. There are some studies in the literature [5, 9, 31] showing that Fitts' law is an effective quantitative method to assess input/output devices including haptic interfaces. Similar to Fitts, Tan [27] proposed that information transfer between a human and a device which is expressed in *bits* is a possible way to make direct performance comparisons between different haptic interfaces. Consequently, the measures of information transfer are better than the measures of accuracy and time in evaluating haptic interfaces.

The resolution limits and other metrics of a haptic device significantly influence the results of human threshold experiments. Weisenberger et al. [32] suggested that if two devices have different fidelities, then it is possible that differences in sensory thresholds might be attributable to the device rather than to the user's perceptual system. Therefore, careful repetition of described human factor studies leads to basic quantitative metrics for the evaluation of haptic interfaces. For instance, sensory threshold values from detection and discrimination experiments for each kind of stimulus can be assessed to evaluate a particular haptic interface by considering the results from other devices and manual threshold experiments with real objects. The resolution and range of a system can also be combined and expressed in terms of the dynamic range. As explained in Sect. 4.5.1, the dynamic range is the ratio of a specified maximum level of a parameter, such as force, to the minimum detectable value of that parameter.

A summary of the proposed performance metrics is shown in Table 6.1. These metrics are complementary to the physical ones summarized in Table 4.2 such as the device workspace, output force, position and force resolution, stiffness, inertia and backdrive friction etc. Overall, they can be used as benchmark indices.

Table 6.2 Psychophysical evaluation testbeds

Experiment		Aim
1	Travel & Selection	To evaluate how well a device supports motor control mode of
2	Selection & Manipulation	haptics
3	Detection	To find force resolution limit of a device by detection experiment
4	Force Discrimination	To find discrimination thresholds affected by a device
5	Texture Discrimination	
6	Size Identification	To evaluate how well a device supports geometric identification of an object
7	Shape Identification	

6.3 Experimental Implementation

Based on the taxonomy of haptic interaction (Figs. 6.2 and 6.3) and corresponding psychophysical experiments in the literature, we have designed and implemented seven testbeds to evaluate effectiveness of a haptic interface. As shown in Table 6.2, these testbeds can be put in four main categories based on the aim of the generic tasks. Three different commercial force-feedback devices were used in order to demonstrate the validity of each testbed.

As shown in Fig. 6.2, there are three generic interaction tasks for the motor control mode. However, only two testbeds may be adequate to evaluate effectiveness of haptic feedback on these three tasks. Since we use the travel term as the navigation of virtual pointer, every selection task requires a travel technique. Similarly, manipulation requires a selection technique [3]. Therefore, we have designed only two experiments for this mode.

6.3.1 Experimental Setup

We have built a universal multi-threaded virtual environment with haptic rendering for different force-feedback devices. As the high-level haptic rendering API provided by the vender of different device varies a lot, we have only implemented the corresponding low-level foundational API provided by the venders, where we can read positions and send forces directly. The graphic rendering is set up with OpenGL and a dual-thread software package is developed to realize the testbeds. The haptic and visual loops were updated at a rate of 500 and 60 Hz, respectively.

The experiments were implemented on a standard laptop with Intel's Core 2 Duo 2.2 GHz CPU. The coordinate system was set in a way that the X-axis points right and the Y-axis points up (see Fig. 6.4). The hardware components of the experimental system include an LCD screen to display the virtual environment, and a force-feedback device to simulate haptic interactions. Throughout all the experiments the tip of the haptic device, which is held by the users, is modeled as a single

Fig. 6.4 Experimental setup
consists of a laptop to run the
designed tasks and to display
the virtual environment, and a
force-feedback device to
simulate haptic interactions.
The *coordinate system* was
set in a way that X-axis
points right and Y-axis points
up

point for collision detection in virtual reality but visually presented to the user as a
small sphere.

6.3.2 Apparatus: 3-DOF Haptic Interfaces

Three commercially available force-feedback devices were tested; PHANTOM
Omni® [26], Xitact™ IHP [21], and omega.3 [8]. Their basic specifications provided by the manufacturers are summarized in Table 6.3. These haptic interfaces
were selected because of their diverse characteristics such as mechanical structure,
output force, application area, cost and workspace. For instance, the omega.3 is a
high-end 3-DOF force feedback device, with high mechanical stiffness, designed
for demanding applications where performance and reliability are critical. On the
other hand, the PHANTOM Omni is a cost-effective 3-DOF haptic device (6-DOF
positional sensing) which has portable design, compact footprint but delivers lower
forces. Finally, the Xitact IHP is a 4-DOF haptic device designed to track the motion
of a surgical instrument and to provide realistic force feedback during the simulation
of a minimally invasive surgical procedure.

Workspaces of these devices are illustrated in Fig. 6.5. As shown in this figure,
the stylus of the PHANTOM Omni was constrained and users held the tip rather than
the stylus during the experiments. Furthermore, the tip of the surgical instrument
was held rather than its handle in the case of the Xitact IHP. In order to avoid any
inconsistency, 3-D translation and rendered force of each device were pre-calibrated
at a uniform scale.

6.3.3 Participants and Procedure

Each device was tested by five graduate students (one woman, four men). Overall
15 users, whose age ranges from 26 to 36, participated in the experiments. They

Table 6.3 Basic specifications according to the manufacturers

	Xitact IHP	PHANTOM Omni	omega.3	
Structure	Hybrid	Serial	Parallel	
Workspace	(r, θ, ϕ) 200, 100°, 100°	(x, y, z) 160, 120, 70	(r, θ, z) 80, 360°, 110	mm
Continuous Force	20	0.88	12	N
Position Resolution	0.057	0.055	0.010	mm
Stiffness	n/a	2.31	14.5	N/mm
Inertia	n/a	45	n/a	g
Backdrive Friction	n/a	0.26	n/a	N

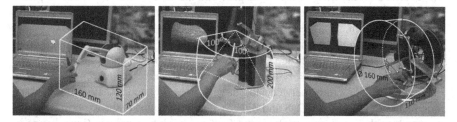

Fig. 6.5 Approximative workspace: (**a**) PHANTOM Omni (~ 0.0013 m^3), (**b**) Xitact IHP (~ 0.0019 m^3), (**c**) omega.3 (~ 0.0022 m^3). *Not to scale*

were all right-handed and had no reported sensory or motor abnormalities. Users were randomly assigned to the devices and each one conducted the complete set of testbeds with the same device in order to avoid any learning effect due to repeated tests. The procedure consisted of a separate training phase and a testing phase for each testbed. In the training phase, participants were told the procedure and they visually and haptically explored every stimuli in the testbed with the corresponding interface. Once they explored the whole testbed in the training phase, the testing phase started. The overall experimental procedure lasted around 90 minutes.

6.4 Psychophysical Testbeds

6.4.1 Experiment 1: Travel and Selection

Fitts' tapping task was chosen for travel and selection testbed which involves selecting targets by touching them and measuring the movement time between tapping the targets. The movement time (*MT*) equation of Fitts' law for one dimensional tasks is given as Shannon formulation [18]

$$MT = a + b \log_2(A/W + 1) \tag{6.1}$$

where a and b are constants determined from experiments by curve fits, A is the distance of movement, and W is the width of target, which corresponds to required accuracy. The logarithmic term is called the *index of difficulty* (*ID*) which is in bits and the reciprocal of b is called the *index of performance* (*IP*) which expresses the results as a performance rate (bit/s). The parameters a and b are system dependent. Thus, they can be used as benchmark metrics for performance comparisons of different haptic devices.

6.4.1.1 Method

In this testbed, users are asked to tap alternately two virtual square plates. The plates are generated on the floor of the workspace and the straight line between the plates is always parallel to the screen. They are modeled as linear springs with a stiffness value of 0.8 N/mm.

In order to sample a wide information range (i.e., $ID =$ from 1.0 to 6.0 bits), seven combinations of amplitude and target size are implemented. For this purpose, three different widths ($W = 2$, 4 and 8 mm) and three center to center distances ($A = 8$, 32 and 128 mm) between the target plates are studied. Two unnecessary combinations (i.e., W, $A = 2$, 8 and 8, 128) are not included to the experimental trials to reduce the number of trials. Force feedback with a visual cue (i.e., change in color) is provided to indicate taps in each case. The order of the 7 trials is randomly assigned. Each trial lasts 15 s and is followed by a 10 s rest period. The number of tabs are recorded.

6.4.1.2 Results & Discussion

Average movement time for one tab in each trial is plotted against *ID* in Fig. 6.6. The data points represent the mean values of five subjects and the vertical bars correspond to the standard deviations over subjects. Significance testing was performed to test the normality of each group of five subjects. First, an F-test was performed between groups for comparing the estimated variances of each trial to test whether the groups come from populations with the same variance. Two-tailed probability results of the F-tests for two given data sets show that the variances of each trial are not significantly different between groups. Then, two-tailed t-tests were performed to assess whether the means of the groups in each trial are statistically different from each other. There was no significant difference of the means between groups.

The *IP* and the intercept (a) were determined by curve fits using Eq. (6.1) as shown in Fig. 6.6. The results of this fundamental task show that the performance metrics are almost the same for these haptic interfaces: the Xitact IHP ($IP = 3.08$, $a = 0.04$), the PHANTOM Omni ($IP = 2.94$, $a = 0.07$), and the omega.3 ($IP = 3.10$, $a = 0.06$). A similar test performed using the PHANTOM Premium 1.5 ($IP = 3.18$, $a = 0.028$) [31] also supports our results. In addition, another study to evaluate 3D stereo haptic workstations [5] provide coherent performance

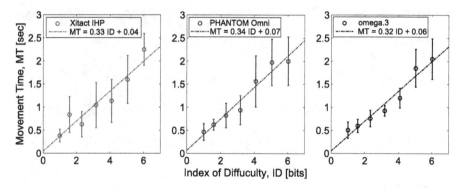

Fig. 6.6 Movement time vs index of difficulty for the travel & selection testbed. The *data points* represent the mean values of five subjects and the *vertical bars* correspond to the standard deviations over subjects. *Dashed lines* represent linear curve fittings to the *data points*. Performance metrics: (*left*) the Xitact IHP ($IP = 3.08$, $a = 0.04$), (*middle*) the PHANTOM Omni ($IP = 2.94$, $a = 0.07$), and (*right*) the omega.3 ($IP = 3.10$, $a = 0.06$)

results for the PHANTOM Premium ($IP = 2$ to 3, $a = 0.03$ to 0.2). The variations in the latter study are due to the different stereo systems used for visualization during the experiments.

These results might be compared with the original Fitts' experiments [7] and other studies on different user input devices by carefully considering many sources of variation in each experiment. Reanalysis of data from the Fitts' experiments using Eq. (6.1) gives $IP = 8.2$, $a = -0.031$ for the experiment with a 1-oz stylus and $IP = 7.2$, $a = -0.070$ with a 1-lb stylus [18]. One might consider $IP = 8.2$ value from this light loaded test (1-oz stylus) as the peak for a simple tapping task by hand. As the movement time increases due to weight of the stylus or any other interference, IP decreases. Therefore, user input devices should support as high as possible IP, which indicates better performance, to be practical and transparent. For example, the pointing tests performed by some user input devices show that the computer mouse has a performance of $IP = 4.5$ [19] which is half of the one with light stylus. Similarly, our result show that the use of haptic devices diminishes the task completion performance of the users down to $IP = 3$. It is also shown that there is no difference between the performance metrics of each device in this basic task.

6.4.2 Experiment 2: Selection and Manipulation

The peg-in-hole test was implemented as a selection and manipulation testbed due to its accuracy and precision in achieving the manipulation task. In this testbed, the user selects the center object (i.e. peg) from a group of objects and places it within a target area (i.e. hole) as shown in Fig. 6.7.

Fig. 6.7 Selection and
manipulation testbed. The
user selects the center object
(*peg*) from a group of objects
and places it within a target
area (*hole*)

6.4.2.1 Method

The cubic objects in Fig. 6.7 including the hole are modeled as rigid bodies on the
floor of the workspace. The hole and the peg are located on a straight line parallel
to the screen. This constraint is imposed to avoid violation of the one dimensional
nature of Fitts' law given in Eq. (6.1). As there are only a small number of rigid
bodies on the scene, the distance between the tip of the haptic device and the cubes
is calculated for collision detection. However, when the peg is selected, the collision
detection is between the walls of the cubic peg and the other rigid bodies. A linear
spring model with a stiffness value of 0.4 N/mm is implemented to render the force
given by the device when there is a collision. The weight of the peg is not considered
in the simulation.

Similar to the tapping task, the *MT* formulation given in Eq. (6.1) is used where
W stands for precision, the difference between the size of hole and peg (i.e., $W =
H - P$). A cube with an edge length (P) of 10 mm is modeled as the peg. In order to
obtain an information range (*ID*) from 3.0 to 8.0 bits, six combinations of distance
and precision are implemented. Four levels of precision ($W = 0.5, 1, 2$ and 4 mm)
and two levels of center to center distances ($A = 32$ and 128 mm) between the
peg and hole are studied. Two unnecessary combinations (i.e., $W, A = 0.5, 32$ and
$4, 128$) are not included in the experimental trials to reduce the number of trials.
Force feedback with a visual cue is provided to indicate selection. The density of
the group is fixed to nine objects and not considered as an experimental variable.
Each trial ends after three successful inserts and there is a 10 s rest period between
two consecutive trials. Task completion time is recorded in this testbed.

6.4.2.2 Results & Discussion

The average movement time for one insert in each trial is plotted against ID in
Fig. 6.8. Similar to the previous section, initial data analysis was performed between
groups for comparing the estimated variances and means of each trial. The results
of the F-tests showed that the variances of some trials are significantly different
between the omega.3 and the others at a level of 0.05. Besides, the two-tailed t-tests
show that the mean values of the Xitact IHP and the PHANTOM in some trials are

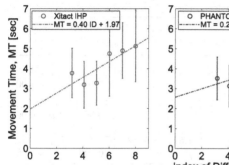
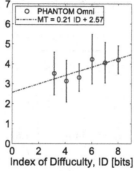
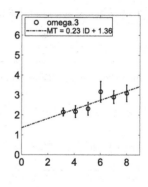

Fig. 6.8 Movement time vs index of difficulty for the selection & manipulation testbed. The *data points* represent the mean values of five subjects and the *vertical bars* correspond to the standard deviations over subjects. *Dashed lines* represent linear curve fittings to the *data points*. Performance metrics: (*left*) the Xitact IHP ($IP = 2.52$, $a = 1.97$), (*middle*) the PHANTOM Omni ($IP = 4.74$, $a = 2.57$), and (*right*) the omega.3 ($IP = 4.37$, $a = 1.36$)

statistically different from the omega.3 at 5 % significance level. This result can be seen graphically comparing the standard deviation in Fig. 6.8.

The *IP* and intercept (*a*) for the selection & manipulation test were determined by curve fits using Eq. (6.1) as shown in Fig. 6.8. The *IP* shows the user performance rate as bit/s. A nonzero intercept *a* is due to the presence of factors at the beginning or ending of a task [18]. Zero intercept implies that a task of zero difficulty takes 0 seconds. Thus, the *a* shows the time spent on the target. As these metrics are system dependent parameters, small *a* and large *IP* values correspond to better user performance due to higher degree of haptic transparency.

Signal transmission between the user and the VE should not be distorted due to losses such as friction and inertia. For instance, comparison of the results show that the omega.3 ($IP = 4.37$, $a = 1.36$) and the PHANTOM Omni ($IP = 4.74$, $a = 2.57$) enable faster movements than the Xitact IHP ($IP = 2.52$, $a = 1.97$). We can conclude that lower inertia and cable transmission of both systems provide improved haptic transparency than the Xitact IHP which is actuated by a friction drive in insertion direction. On the other hand, the small difference in the index of performance between the omega.3 and PHANTOM Omni means that the kinematics and mechanical structure of the both devices provide the same user performance at manipulation tasks. However, the difference in *a* values shows that haptic feedback provided by the omega.3 at the selection and releasing of the peg is more appropriate than the PHANTOM Omni. This also explains the reduced deviation of movement time over subjects when the omega.3 is used.

6.4.3 Experiment 3: Detection

Absolute thresholds for position, velocity, force, pressure, stiffness, and viscosity can be used as quantitative performance metrics of a haptic interface (Fig. 6.3). In

Table 6.4 ANOVA results of the force detection experiment

	Factors		Xitact IHP	PHANTOM Omni	omega.3
Significant	Force Stimulus	$F(5, 140)$	7.84, $p < 0.001$	28.75, $p < 0.001$	23.84, $p < 0.001$
	Direction	$F(5, 140)$	19.50, $p < 0.001$	3.76, $p < 0.005$	7.85, $p < 0.001$
Insignificant	User	$F(4, 140)$	1.66, $p > 0.100$	0.97, $p > 0.100$	2.32, $p > 0.050$

this testbed, a force detection experiment is examined revealing the absolute threshold for force. Then, this value is compared with the human hand's threshold sensitivity to force on the fingertips which is 0.06 N as shown in Table 3.2.

6.4.3.1 Method

The method of constant stimuli is chosen for this testbed. Force stimuli ranging from 0.1 to 0.6 N with 0.1 N increments are fully crossed with the three axis (X, Y and Z) and two directions ($+$ and $-$). Each trial lasts 10 s and a force stimulus with different magnitude and direction is randomly presented to the user after 4 s. At the end of each trial, users response whether the stimulus is detectable or not.

6.4.3.2 Results & Discussion

A linear statistical model was chosen for the analysis of detection testbed results. An analysis of variance (ANOVA) with factors of stimulus level (six), direction (six), user (five) and interaction between stimulus and direction was performed to check the validity of the model for each device. Statistical results of this analysis are presented in Table 6.4. ANOVA revealed that only the stimulus and direction were statistically significant for every device. The factor "user" refers to the differences among the subjects in each group. The variation within the subjects are not statistically significant for any group.

The minimum detectable force stimulus for each direction was calculated based on the coefficients of the linear model and 50 % correct response. Absolute force thresholds for each device and direction are shown in Table 6.5. These values can be interpreted as the minimum force level that can be rendered by the interfaces on the specified axis. These interesting results provide useful information about the limits of the devices that cannot be retrieved easily from the known specifications. For example, the PHANTOM Omni can render very low forces (i.e. 0.2 N) on X and Z axes as its extension structure is lighter than the other two devices. Yet, lack of gravity compensation prevents the same performance for up-down direction. The

Table 6.5 Quantitative results for the force detection testbed

Performance Metric	Axis	Xitact IHP	PHANTOM Omni	omega.3	Unit
Absolute Threshold	X (+, −)	0.6, 0.6	0.3, 0.2	0.3, 0.2	N
	Y (+, −)	0.3, 0.2	0.4, 0.5	0.3, 0.2	N
	Z (+, −)	0.4, 0.5	0.2, 0.3	0.5, 0.4	N
Dynamic Range	X	19	13	36	dB
	Y	40	7	36	dB
	Z	20	13	30	dB

minimum force that can be rendered on this axis is 0.4 N which is much higher than human hand's sensitivity to force on the fingertips. It is also obvious from Table 6.5 that the Xitact IHP cannot generate low forces less than 0.6 N on the left-right axis. Possible reason for this limitation is that the transmission ratio is higher on this axis than the others due to the special kinematics design. The kinematics results in higher inertia that is felt at the tip while moving it left and right. Thus, the absolute force threshold increases for this direction. Results also show that, although the omega.3 has low force sensitivity on X and Y axes because of the effective gravity and friction compensation, its parallel mechanism does not allow to render lower forces on the front-back axis.

In addition to the force threshold, dynamic range, ratio of the maximum force to the minimum detectable force, was also calculated according to the maximum continuous force specified by the manufacturers. As shown in Table 6.5, the PHANTOM Omni has considerably narrow force rendering range (7 dB) in up-down direction which is 1/5 of the omega.3 (36 dB) and the Xitact IHP (40 dB).

6.4.4 Experiment 4: Force Discrimination

Differential threshold values can be determined for a specific haptic interface by performing a discrimination experiment. Weber fractions for human perception are given in Table 3.2. A discrimination testbed indicating force thresholds was designed considering the Weber fraction for force (0.07).

6.4.4.1 Method

Two virtual constant force regions, one of which is being the reference, are presented side by side in the middle of the workspace (see Fig. 6.9). Two force fields have different magnitudes and they are symmetric about the YZ-plane. Users are asked to explore both regions and determine whether the other force stimulus is different than the reference or not.

Fig. 6.9 Force discrimination testbed. Users are asked to explore two virtual force regions and determine whether the stimuli are different or not in terms of force magnitude

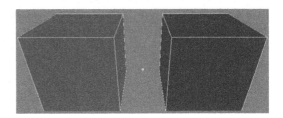

Three factors that must be controlled in this discrimination experiment are reference stimulus (I), Weber fraction ($\Delta I / I$) and direction of the force stimulus. In order to treat these three factors at the same time with less trials, a 6 × 6 Latin square design is implemented. The experimental design is shown in Table 6.6. Equally spaced values of I (from 1 to 6 N) are given in rows and columns represent the six values of $\Delta I / I$. The directions ($+X, -X, +Y, -Y, +Z$ and $-Z$) are assigned randomly within rows and columns with each direction appearing once in every row and every column. A total of 36 combinations are presented to the user in a random order.

Since the peak force is 3.3 N for the PHANTOM Omni, the force stimulus presented to the user were decreased by half (i.e., I from 0.5 to 3 N) during the experiments performed with the PHANTOM Omni.

6.4.4.2 Results & Discussion

An ANOVA with factors of stimulus level (six), direction (six), weber fractions (six), and user (five) was performed. The significant factors vary for each device as shown in Table 6.7. The only common significant factor among the devices is the weber fraction term.

Taking the 75 % correct response point as the differential threshold, force weber fractions were obtained for each axes (see Table 6.8). The X-axis results were excluded from the table because careful investigation of the results reveals that the experiments performed along the X-axis were erroneous. In fact, during the user tests it was also observed that the users had difficulties to discriminate forces given

Table 6.6 The 6 × 6 Latin square design

Reference, $I(N)$	Weber Fractions, $\Delta I / I$					
	0.1	0.2	0.3	0.4	0.5	0.6
1	$+X$	$-X$	$+Y$	$-Y$	$+Z$	$-Z$
2	$+Y$	$-Y$	$+Z$	$-Z$	$+X$	$-X$
3	$-Z$	$+X$	$-X$	$+Y$	$-Y$	$+Z$
4	$-Y$	$+Z$	$-Z$	$+X$	$-X$	$+Y$
5	$-X$	$+Y$	$-Y$	$+Z$	$-Z$	$+X$
6	$+Z$	$-Z$	$+X$	$-X$	$+Y$	$-Y$

Table 6.7 ANOVA results of the force discrimination experiment

	Factors		Xitact IHP	PHANTOM Omni	omega.3
Significant	Weber Fractions	$F(5, 160)$	3.26, $p < 0.010$	3.92, $p < 0.005$	9.60, $p < 0.001$
	Force Stimulus	$F(5, 160)$	n/a	n/a	3.62, $p < 0.005$
Insignificant	User	$F(4, 160)$	2.04, $p > 0.050$	2.40, $p > 0.050$	1.80, $p > 0.100$

Table 6.8 Weber fractions for the discrimination testbeds

Variable	Axis	Xitact IHP	PHANTOM Omni	omega.3
Force	Y	0.5	0.4	0.3
	Z	0.4	0.5	0.3
Texture	Z	0.5	0.4	0.4

along the X-axis. This can be explained by the fact that different arm muscles were used against the X-axis forces in the left-right force fields since they were symmetric about YZ-plane in this experiments.

The weber fractions given in Table 6.8 are highly related to the force resolution and transparency of the haptic interfaces (i.e., the smaller the value, the smaller the losses in force transmitted to the user). Perfect transparent case is when force weber fraction is 0.07 which is the human discrimination threshold for force. Considering this fact, we can conclude that the omega.3 (force weber fraction of 0.3) has better force resolution and haptic transparency than the others. However, even this low threshold is four times the value coming from the manual discrimination tests. Therefore, any experimenter performing psychophysical studies by any haptic interface should be aware of the device limitations and capabilities. Otherwise, the results of such a study will be misleading.

6.4.5 Experiment 5: Texture Discrimination

The most common way to measure humans' performance in haptic discrimination of textured surfaces is to use periodic gratings. For instance, psychophysical studies using 2D raised-dot patterns revealed that humans could distinguish periodic surfaces in which the period of the dots differed by only 2 % [16] (also shown in Table 3.3). Similarly, an easy way to assess the texture rendering capability of haptic interfaces is to use different periodic gratings and ask users to distinguish them. Therefore, a texture discrimination testbed was designed to measure the threshold

Fig. 6.10 Texture
discrimination testbed. Users
are asked to explore two
virtual gratings and indicate
whether they may distinguish
the difference between the
periods of gratings or not.
Only a graphical cursor
indicating the current location
is displayed as visual
feedback during the trials

of the spatial period. As in the force discrimination testbed, the method of constant
stimuli was implemented.

6.4.5.1 Method

The proposed methodology consists of two textures that are modeled as sinusoidal
gratings located on the side walls of the virtual environment facing each other (see
Fig. 6.10). A linear spring model with a stiffness value of 0.5 N/mm is implemented
to render the textures. Collision detection between the cursor and the textures is
based on the analytical representation of the textures.

The experimental design is based on the Latin square as shown in Table 6.6.
Three independent variables are the amplitude and spatial period of gratings and
the Weber fractions for period. The amplitude range goes from 0.5 to 3 mm hav-
ing 0.5 mm increment. The spatial period range goes from 1 to 6 mm with 1 mm
increment. The weber fractions take the values of 0.05, 0.1, 0.2, 0.3, 0.4 and 0.5.
The orientation of the gratings is kept constant. Textures have different periods, but
the same amplitude. Users are asked to explore them indicating whether they may
distinguish the difference between the periods of gratings or not. The amplitude of
the gratings is varied for different pairs of textures. A graphical cursor indicating the
current location of the tip of the device is displayed. Users are instructed to follow a
straight line along the orientation and keep their scanning velocity as slow as possible and constant (i.e., less than 10 cm/s). These requirements are imposed in order
to simulate the gratings accurately according to the scanning velocity limit defined
by Campion and Hayward [4].

6.4.5.2 Results & Discussion

As shown in Table 6.9, the ANOVA results of the texture discrimination experiment
show that the effect of the weber fraction was the most significant factor for all
devices considered in this work. The amplitude of gratings was also significant for
the omega.3. Analysis of the results show that the gratings with amplitudes less than
1.0 mm were not distinguishable by this device. Similarly, the spatial difference

Table 6.9 ANOVA results of the texture discrimination experiment

	Factors		Xitact IHP	PHANTOM Omni	omega.3
Significant	Weber Fractions	$F(5, 160)$	17.79, $p < 0.001$	18.16, $p < 0.001$	12.24, $p < 0.001$
	Amplitude	$F(5, 160)$	n/a	n/a	6.19, $p < 0.001$
Insignificant	User	$F(4, 160)$	0.31, $p > 0.5$	1.96, $p > 0.1$	0.55, $p > 0.5$

between the gratings of 0.5 mm amplitude was not recognizable by the PHANTOM Omni, although the amplitude level is not statistically significant for this device. Using the coefficients of the linear model and 75 % correct response, the weber fraction of the distinguishable spatial period is found to be 0.5 for the Xitact IHP and 0.4 for the PHANTOM Omni and the omega.3 as shown in Table 6.8. These values are much higher than the haptic discrimination threshold of humans when touching real textured surfaces which is only 0.02. This result is unlikely to be due to the position resolution of the devices. The low speed reconstruction limit [4], which takes into account the device resolution, gives us that the PHANTOM Omni and the Xitact IHP are able to regenerate periodic gratings with 0.5 mm pitch. As for the omega.3, even a finer pitch of 0.1 mm is possible. Possible reason of higher threshold is the lack of high frequency information which is known to be crucial in texture identification. The users were instructed to scan slowly to prevent high frequency vibrations. Even though vibrations occur, these force feedback devices act like a lowpass filter due to their limited mechanical bandwidth. Therefore, it was difficult to discriminate fine surface textures by these devices.

6.4.6 Experiments 6 & 7: Size & Shape Identification

The identification of geometric properties can be divided into two; (1) shape and (2) size identification. Although these two subtasks require different experiments, similar quantitative performance metrics can be used for evaluation. Since humans can correctly identify 4 sphere sizes ranging from 10 to 80 mm in radius [27], 2 bits of information transfer (IT) can be used as the threshold of identification performance for device evaluation. In other words, humans can correctly identify up to 4 stimulus categories ($k = 2^{IT}$) which should be supported by each haptic interface.

In the absolute identification paradigm [27], given k stimuli, S_i, corresponds to k responses, R_j , where $1 \leq i, j \leq k$. The maximum likelihood estimate of information transfer, IT_{est}, for a particular response pair (S_i, R_j) is given as

$$IT_{est} = \sum_{j=1}^{k} \sum_{i=1}^{k} \frac{n_{ij}}{n} \log_2 \left(\frac{n_{ij}.n}{n_i.n_j} \right) \tag{6.2}$$

Fig. 6.11 Shape
identification testbed. Users
are asked to identify one of
the quadric shapes without
any visual feedback. An
asymmetric saddle is shown
here for demonstration
purpose

where n is the total number of trials, n_{ij} is the number of joint event (S_i, R_j) occurs,
and $n_{i,j} = \sum_{i,j=1}^{k} n_{ij}$.

6.4.6.1 Method: Size identification

The size identification testbed is based on the experiment conducted by Tan [27].
It involves haptically identifying different sizes of objects which are presented in a
random order. A virtual sphere is displayed in the middle of the workspace without a
visual cue. A linear spring model with a stiffness value of 0.4 N/mm is implemented
to render the sphere. Four different sizes (diameter, $D = 10, 30, 60$ and 80 mm) are
considered in order to obtain 2 bits information range. At each trial, users haptically explore the sphere and indicate the size by pressing the corresponding key on
the keyboard. To avoid over-correct results, the total number of trials is set to 100
($n > 5k^2$) as proposed by Tan [27].

6.4.6.2 Method: Shape Identification

The method proposed by Kirkpatrick and Douglas [14] is modified and used as
shape identification testbed. The user has to identify one of the quadric shapes.
These shapes are constructed from two orthogonal parabolas and distinguished
from one to another by the shape index S, a quantity describing the shape [13].
Four quadratic shapes are modeled: concave and convex cylindrical paraboloids
($S = -0.5$ and 0.5 respectively) and two asymmetric saddles ($S = -0.25$ and 0.25).
One of the quadratic shapes ($S = -0.5$) is shown in Fig. 6.11. They are presented
in the center of the workspace without a visual cue. Collision detection between the
cursor and the shape surface is implemented analytically using the quadratic equations which describe the shapes. The penetration depth is calculated based on the

Table 6.10 Information transfer results from the identification experiments

Testbed	Xitact IHP	PHANTOM Omni	omega.3	Unit
Size	1.36	1.36	1.74	bits
Shape	1.12	1.30	1.65	bits

equations and used for force feedback with a stiffness value of 0.4 N/mm. A similar experimental procedure as in the size identification test is performed. Curvedness, location and orientation of the shapes are not altered.

6.4.6.3 Results & Discussion

The *IT* of identification performance for size and shape was calculated using Eq. (6.2). The estimated information transmitted in the size and shape identification testbeds are given in Table 6.10. These values are less than 2 bits of *IT*, the threshold identification performance of human, showing that the performance is degraded by the use of the haptic interfaces. Nevertheless, the omega.3 considerably supports the users' accomplishments in categorizing different size and shapes up to 4 stimulus. Even though, the differences in *IT* values could be explained by haptic rendering quality of each device, this might also be attributed to the device kinematics. Obviously, the surgery specific kinematic design of the Xitact IHP impedes proper exploration and geometric identification of objects. On the other hand, the parallel kinematics design of the omega.3 has a decoupled hand rotation which is always parallel to the base. This provides a more natural exploration possibility.

6.5 General Discussion

Resulting quantitative performance metrics of the testbeds are summarized in Table 6.11. These metrics are highly related to the basic characteristics of the haptic interfaces such as transparency, position and force resolution, stiffness and kinematics. Any differences in these values can be attributed to the one or more device characteristics. They also provide additional information about the interfaces, which is not easily derived from the given specifications, since detailed information on characteristics of such devices are not always provided by manufacturers (see Table 6.3 for the available basic specifications of the interfaces). Besides, since the experiments are performed in the devices' normal operating conditions throughout their workspace, the performance results are not biased to the best performing point. Yet, the results are randomized by the movement of the users over the workspace.

The main advantage of these metrics is the straightforwardness to link them to the rendering quality perceived by the user. For a user without a technical background, results of the physical evaluation are not straightforward to interpret them in terms of target application. On the other hand, the psychophysical testbeds, which are categorized based on the haptics mode as shown in Table 6.2, are intuitively connected to the aim of a task. Depending on the target application, users may decide which testbed and metrics they should consider for device selection. For instance, if the target application involves rendering objects with surface details, users may refer to a device's performance on the texture discrimination and geometric identification testbeds.

These metrics also provide a deeper understanding of the sensory resolution connected to a device. Comparison of these values with those from the human psychophysical experiments with real objects leads to a better understanding of the influence of haptic interfaces. Considering human perceptual thresholds as a norm for meaningful device comparison and assessment, the metrics provide device-specific limits. Moreover, they also quantitatively show how much differently the devices perform in a given task. Therefore, these tests help us to determine actual or probable usefulness of a device with quantifiable characteristics.

Transparency is an important performance characteristic of haptic interfaces. However it is difficult to quantify because of effects such as friction, inertia, mechanical backlash as demonstrated in Chap. 4. Such effects are not readily quantified in specifications and are difficult to model. On the other hand, the results of the first two experiments are highly related to the haptic transparency. Therefore, we suggest that the index of performance *IP* and the intercept *a* (i.e., the time spent on the target) which are calculated using Fitts' law could be considered as one set of quantitative measures of transparency. Our results support this suggestion. One might conclude that the haptic transparency of all the devices are the same since there is no significant difference between the results of the first testbed. However, this might also mean that the travel and selection task, which is Fitts' tapping task, alone is not sufficient to reveal the haptic device characteristics as it does not include any task requiring haptic accuracy. Moreover, the more sophisticated task in the selection and manipulation testbed serves as a complimentary experiment to this one. The *IP* and the intercept *a* values from this experiment vary for each haptic interface. Lower inertia and cable transmission of the PHANTOM Omni and the omega.3 provide higher haptic transparency. Although a friction and inertia compensation algorithm is implemented in the Xitact IHP, the friction drive actuation system at the insertion still shows up as a loss. Besides, the differences in *a* values also show that the force feedback provided by the omega.3 during the pick and place is more proper to accomplish the peg-in-hole task. After all, the *IP* values are still less than the ones from the actual Fitts' experiments performed using 1-oz stylus ($IP = 7.2$) as shown in Sect. 6.4.1. Thus, new user interface designs should aim for higher *IP* indicating better transparency and performance.

Absolute force thresholds in Table 6.11 are the minimum force that can be rendered by the interface on the specified axis. From these results including the dynamic range, we can draw some useful conclusions on force resolution, friction and inertia in the system for each direction. For example, the PHANTOM Omni can render very low forces (i.e. 0.2 N) due to its lighter structure. However, the absolute force threshold for the up-down direction is double of this value due to lack of a

Table 6.11 Quantitative results of the testbeds

Testbed	Performance Metric	Axis	Haptic Interface			Unit
			Xitact IHP	PHANTOM	omega.3	
Travel & Selection	Index of Performance (*IP*)		3.08	2.94	3.10	bit/s
	Intercept (*a*)		0.04	0.07	0.06	s
Selection & Manipulation	Index of Performance (*IP*)		2.52	4.74	4.37	bit/s
	Intercept (*a*)		1.97	2.57	1.36	s
Detection Force	Absolute Threshold	X (+, −)	0.6, 0.6	0.3, 0.2	0.3, 0.2	N
		Y (+, −)	0.3, 0.2	0.4, 0.5	0.3, 0.2	N
		Z (+, −)	0.4, 0.5	0.2, 0.3	0.5, 0.4	N
	Dynamic Range	X	19	13	36	dB
		Y	40	7	36	dB
		Z	20	13	30	dB
Discrimination Force	Weber Fraction	Y	0.5	0.4	0.3	
		Z	0.4	0.5	0.3	
Texture		Z	0.5	0.4	0.4	
Identification	Information Transfer (*IT*)					
Size			1.36	1.36	1.74	bits
Shape			1.12	1.30	1.65	bits

gravity compensation. On the other hand, it is not possible to feel low forces on the front-back axis compared to other directions with the omega.3. The minimum force that can be rendered by these devices is still much higher than the absolute force threshold of humans on the fingertips which is 0.06 N. In addition to these results, comparing the weber fractions from the discrimination experiments to the human discrimination threshold for force, 0.07, we can determine a level for the haptic transparency. For instance, the omega.3 (force weber fraction of 0.3) has better haptic transparency than the others. Nonetheless, this threshold is four times higher than the human capability of discriminating force. In a similar manner, the texture discrimination values are much higher than the humans' performance in haptic discrimination of periodic gratings which is as small as 0.02.

The information transfer values are also useful benchmark metrics for device comparison. As shown in Table 6.11, the users did not have the same identification performance as with real objects (2 bits). This shows that the performance is degraded by the use of the haptic interface. The differences in IT values could be explained by the device kinematics. While the kinematic design of the Xitact IHP impedes geometric identification of objects, the parallel kinematics design of the omega.3 provides more natural exploration.

The haptic loop was set to a lower value (500 Hz) than the normally preferred update rate (1 kHz) throughout all the experiments because the Xitact IHP cannot support higher rates due to latencies introduced by the USB communication. We have implemented the same update rate for all devices to unify the virtual environment and obtain comparable results. The update rate is highly related to the stability of the contact with stiff objects. Yet, as the maximum stiffness used in our experiments was 0.8 N/mm, which was tuned to obtain passive contact, 500 Hz is indeed acceptable [20]. This value is also enough to render sharp edges. Nevertheless, setting the device controllers to run lower than what they can support might be seen as downgrading the performance of the two high-end devices (the PHANTOM Omni and the omega.3 normally run at 1 kHz). In fact, this is a right concern. Since the maximum stiffness is also a performance metric for haptic interfaces, a better comparison could be obtained when the simulations are run at each device's optimal operating conditions. In other words, while using the same algorithm for VE generation, different haptic update rates and stiffness values adapted for each device should be used.

Statistical analysis showed that the experimental results did not significantly depend on the subjects but any other factor tested. Results clearly demonstrate that the variations within the subjects were not statistically significant ($p > 0.05$) for all the experiments. On the other hand, there was at least one highly significant effect ($p < 0.005$) for each experiment. In addition, since the assignment of subjects to the devices was randomized, any significant effect observed between the groups was not a characteristic of the individuals in the group and could be linked to the device type. Nevertheless, even if the ANOVA did not reveal particular between-subject variance, five subjects per device might not be great enough to infer statistical results. Therefore, the number of subjects and repetitions should be increased [33].

Moreover, overall analysis shows significant differences between our experiments and the human psychophysical tests with real objects. This clearly demonstrates the limits of haptic interfaces and virtual environment and the importance of high quality for haptic interfaces. Considering the increased number of psychophysical tests performed by commercial haptic interfaces, any experimenter performing this kind of studies must be well aware of the device limitations and capabilities. These limits should be quantified in the frame of a standardized method. Otherwise, the results of such a psychophysical study might be misleading.

6.6 Conclusions

In order to find sensory thresholds associated with a haptic interface and also to evaluate how well this interface supports haptic interaction, we have implemented seven testbeds. The testbeds were applied to three force-feedback devices and detailed comparison of the results (with the human perceptual thresholds as well) were provided.

During the testbed experiments, other performance metrics such as the force and velocity profile and used workspace could also be collected for interface evaluations. Quantitative metrics are suitable for validation, yet, they do not completely represent the usability of the interface. Thus, qualitative metrics based on users' experiences during testbeds should be collected. User fatigue, discomfort and lack of immersion are strongly influenced by improper use of haptic feedback.

One of the potential limiting factors of generic evaluations is the assumption of a "perfect" virtual reality environment. Results of these kind of experiments depend on the rendering technique. However, in order to make a comparison between different interfaces, a standard virtual environment system should be used. Therefore, we have used the same VE experimental setup for all interfaces to work out clearly the differences only due to the devices. All the other factors were controlled; thus the experimental results show a fair comparison of three devices for basic haptic interaction tasks. In the literature, one can find similar studies for other devices. However, since there is no across-study consensus, it will be misleading to compare the results. Therefore, we believe that our unified approach covering all aspects of haptic interaction for three devices is a considerable contribution to the state of the art. In addition, we have paid careful attention to the design of experiments to identify the exhaustive variables with a minimum number of experiments but still statistically appropriate way.

Although the testbeds were tested only on force-feedback devices, the taxonomy and the basic principles of the testbeds can also be applied to tactile feedback devices to some extent. However, the thermal mode of haptic interaction is not included in the taxonomy. The next chapter describes the application of the testbeds to a tactile feedback device.

References

1. Avizzano, C.A., Solis, J., Frisoli, A., Bergamasco, M.: Motor learning skill experiments using haptic interface capabilities. In: Proc. of 11th IEEE International Workshop on Robot and Human Interactive Communication, pp. 198–203 (2002)
2. Basdogan, C., Ho, C., Srinivasan, M.A., Slater, M.: An experimental study on the role of touch in shared virtual environments. ACM Trans. Comput.-Hum. Interact. 7(4), 443–460 (2000)
3. Bowman, D.A., Hodges, L.F.: Formalizing the design, evaluation, and application of interaction techniques for immersive virtual environments. J. Vis. Lang. Comput. 10(1), 37–53 (1999)
4. Campion, G., Hayward, V.: Fundamental limits in the rendering of virtual haptic textures. In: World Haptics Conference, pp. 263–270 (2005)
5. Chun, K., Verplank, B., Barbagli, F., Salisbury, K.: Evaluating haptics and 3d stereo displays using Fitts' law. In: Proc. of the 3rd IEEE Workshop on HAVE, pp. 53–58 (2004)
6. Durlach, N.I., Delhorne, L.A., Wong, A., Ko, W.Y., Rabinowitz, W.M., Hollerbach, J.: Manual discrimination and identification of length by the finger-span method. Percept. Psychophys. 46(1), 29–38 (1989)
7. Fitts, P.M.: The information capacity of the human motor system in controlling the amplitude of movement. J. Exp. Psychol. 47, 381–391 (1954)
8. Force Dimension: Omega. http://www.forcedimension.com/ (2012)
9. Hannaford, B., Wood, L., McAffee, D., Zak, H.: Performance evaluation of a six axis generalized force reflecting teleoperator. IEEE Trans. Syst. Man Cybern. 21, 620–633 (1991)
10. Harders, M., Barlit, A., Akahane, K., Sato, M., Szkely, G.: Comparing 6dof haptic interfaces for application in 3d assembly tasks. In: Proc. of Eurohaptics'06 (2006)
11. Jandura, L., Srinivasan, M.A.: Experiments on human performance in torque discrimination and control. Dyn. Control Syst., ASME 55-1, 369–375 (1994)
12. Jones, L.A., Lederman, S.J.: Human Hand Function. Oxford University Press, London (2006)
13. Kappers, A.M., Koenderink, J.J., Lichtenegger, I.: Haptic identification of curved surfaces. Percept. Psychophys. 56(1), 53–61 (1994)
14. Kirkpatrick, A.E., Douglas, S.A.: Application-based evaluation of haptic interfaces. In: Proc. of the 10th Haptic Symposium, p. 32 (2002)
15. Klatzky, R.L., Loomis, J.M., Lederman, S.J., Wake, H., Fujita, N.: Haptic identification of objects and their depictions. Percept. Psychophys. 54(2), 170–178 (1993)
16. Lamb, G.D.: Tactile discrimination of textured surfaces: psychophysical performance measurements in humans. J. Physiol. 338(1), 551–565 (1983)
17. Lawrence, D.A., Pao, L.Y., Dougherty, A.M., Salada, M.A., Pavlou, Y.: Rate-hardness: a new performance metric for haptic interfaces. IEEE Trans. Robot. Autom. 16(4), 357–371 (2000)
18. MacKenzie, I.S.: Fitts' law as a research and design tool in human-computer interaction. Hum.-Comput. Interact. 7, 91–139 (1992)
19. MacKenzie, I.S., Sellen, A., Buxton, W.: A comparison of input devices in elemental pointing and dragging tasks. In: Proc. of the CHI'91 Conference on Human Factors in Computing Systems, pp. 161–166 (1991)
20. MacLean, K.E., Snibbe, S.S.: An architecture for haptic control of media. In: Proc. of the 8th Ann. Symp. on Haptic Interfaces for Virtual Environment and Teleoperator Systems, ASME/IMECE, pp. 3–5 (1999)
21. Mentice SA (formerly Xitact SA): Xitact IHP. http://www.mentice.com/ (2012)
22. Murray, A.M., Klatzky, R.L., Khosla, P.K.: Psychophysical characterization and testbed validation of a wearable vibrotactile glove for telemanipulation. Presence: Teleoperators Virtual Environ. 12(2), 156–182 (2003)
23. Oakley, I., McGee, M.R., Brewster, S.A., Gray, P.D.: Putting the feel in 'look and feel'. CHI, pp. 415–422 (2000)
24. O'Malley, M., Goldfarb, M.: The effect of force saturation on the haptic perception of detail. IEEE/ASME Trans. Mechatron. 7, 280–288 (2002)

25. Samur, E., Wang, F., Spaelter, U., Bleuler, H.: Generic and systematic evaluation of haptic interfaces based on testbeds. In: Proc. of IEEE/RSJ Int. Conf. on Intelligent Robots and Systems, IROS'07 (2007)
26. Sensable Technologies, Inc.®: PHANTOM Omni®. http://www.sensable.com/ (2012)
27. Tan, H.: Identification of sphere size using the phantom: towards a set of building blocks for rendering haptic environment. In: ASME Annual Meeting, pp. 197–203 (1997)
28. Turk, M., Robertson, G.: Perceptual user interfaces. Commun. ACM **43**(3), 32–34 (2000)
29. Unger, B.J., Nicolaidis, A., Berkelman, P.J., Thompson, A., Klatzky, R.L., Hollis, R.L.: Comparison of 3-d haptic peg-in-hole tasks in real and virtual environments. In: IEEE/RSJ, IROS, pp. 1751–1756 (2001)
30. Wall, S.A., Harwin, W.: A high bandwidth interface for haptic human computer interaction. Mechatronics **11**, 371–387 (2001)
31. Wall, S.A., Harwin, W.S.: Quantification of the effects of haptic feedback during a motor skills task in a simulated environment. In: Proc. of the 2nd PHANToM Users Research Symposium, pp. 61–69 (2000)
32. Weisenberger, J., Kreier, M., Rinker, M.: Judging the orientation of sinusoidal and square-wave virtual gratings presented via 2-dof and 3-dof haptic interfaces. Haptics-e **1**(4) (2000)
33. Wichmann, F.A., Hill, N.J.: The psychometric function: I. Fitting, sampling, and goodness of fit. Percept. Psychophys. **63**(8), 1293–1313 (2001)

Chapter 7
Application to a Tactile Display

Abstract This chapter explores the haptic rendering capabilities of a variable friction tactile interface through psychophysical experiments. In order to obtain a better understanding of the sensory resolution associated with the Tactile Pattern Display (TPaD), friction discrimination experiments are conducted. During the experiments, subjects are asked to explore the glass surface of the TPaD using their bare index fingers, to feel the friction on the surface, and to compare the slipperiness of two stimuli displayed in sequential order. The fingertip position data is collected by an infrared frame and normal and translational forces applied by the finger are measured by force sensors attached to the TPaD. The recorded data is used to calculate the coefficient of friction between the fingertip and the TPaD. The experiments determine the just noticeable difference (JND) of friction coefficient for humans interacting with the TPaD. This chapter is based on Samur et al. (Proc. of Human Vision and Electronic Imaging XIV, vol. 7240, 2009).

7.1 Tactile Pattern Display (TPaD)

The Tactile Pattern Display (TPaD) has been developed at the Laboratory of Intelligent Mechatronics Systems of Northwestern University [13]. The TPaD employs ultrasonic vibrations to create a squeeze film of air between the vibrating surface and a fingertip, thereby reducing the friction. These vibrations are far higher than the vibration perception of human at the finger tip. Similar devices have been developed by Watanabe and Fukui [12], Nara et al. [8], and Biet et al. [1, 3]. The variation of the friction level creates shear forces on the fingertip, which are used to present both geometric and material properties of an object through the EPs of lateral motion. More recently, two new tactile devices have been developed at Northwestern University based on the same surface haptics phenomenon: ShiverPaD [4] and Large Area TPaD [7].

Chapter 7 is published with kind permission of © SPIE 2009. All rights reserved. Originally published as: Evren Samur, J. Edward Colgate and Michael A. Peshkin, "Psychophysical evaluation of a variable friction tactile interface", Proc. SPIE 7240, 72400J (2009); http://dx.doi.org/10.1117/12.817170.

Fig. 7.1 Rectangular TPaD:
Two 16 mm diameter, 0.5 mm
thick piezo ceramic disks
glued to the top and bottom of
a 50.8 × 25.4 × 4.9 mm glass
plate (*left*). The plate is fixed
to the aluminum frame at
nodal lines by 4 nylon tip set
screws (*right*)

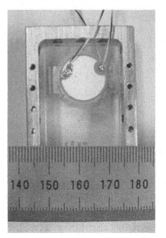
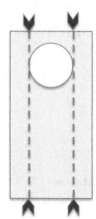

Psychophysical evaluation is necessary for several reasons. First, we want to obtain a deeper understanding of the sensory resolution associated with the TPaD. Perhaps because such devices were not previously available, there is a dearth of data describing detection thresholds and subjective magnitude associated with friction and patterns created by variable friction. Second, we want to understand how variations in friction can give rise to the percept of shape. Anecdotal evidence assures us that shape recognition can be driven by a frictional display, but the mechanisms are not understood.

From a perceptual perspective, friction has received very little psychophysical analysis. The lack of systematic work on the perception of friction reflects the difficulty of controlling a range of friction values in physical objects. Virtual environments present a potential solution to this problem, but studies of friction perception with force-feedback devices have not been directed toward an understanding of basic human processes. Notably, these studies required subjects to experience friction forces via a hand-held probe, not via the bare fingertip. The TPaD is unique in that the friction-controlled virtual environment is experienced by the bare fingertip. In this work, we will exploit not only the TPaD's ability to control friction, but its ability to modulate friction spatially and temporally, producing lateral force fields (LFFs). Robles-De-La-Torre and Hayward [9] found that, despite the loss of all proprioceptive and kinesthetic geometric cues, subjects were able to identify virtual bumps and holes given the appropriate lateral force fields. When subjects were given the physical displacement of a bump but played the LFF of a hole, the subjects ignored the geometric cues and identified the object as a hole.

7.2 Physical Evaluation

This display consists of two 16 mm diameter, 0.5 mm thick piezo ceramic disks
(PI Ceramic GmbH) glued to the top and bottom of a 50.8 × 25.4 × 4.9 mm glass

Fig. 7.2 Transfer function of the TPaD including the amplifier. White noise voltage is given by the signal generator and velocity of vibrations measured by the LDV

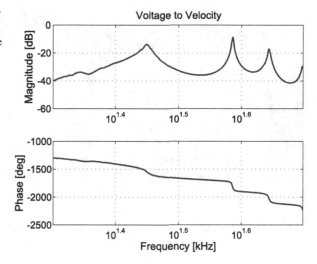

plate (see Fig. 7.1). The plate is fixed to an aluminum frame by 4 nylon-tip set screws. The disks generate ultrasonic vibrations on the plate when an alternating voltage difference is applied. A high frequency, sinusoidal signal is generated by a signal generator and multiplied by a computer generated analog output to be able to control the amplitude of the excitation voltage. Then, the signal is amplified and applied to the piezo disks.

In order to characterize the frequency behavior of the TPaD, we have performed a dynamic analysis. The velocity of the vibrations was measured using a Laser Doppler Vibrometer (LDV) as the piezo disks were driven with white noise. These data were used to derive a transfer function, which is shown in Fig. 7.2. A resonant peak is found at 38.5 kHz, and this peak has a Q-factor (resonant frequency divided by the peak width at half-height) of 100. Flexural vibrations at this ultrasonic frequency create a squeeze film of air between the plate surface and the finger, thus reducing the friction between the plate and the fingertip as the amplitude of the applied voltage is increased. The mechanisms of friction reduction have been discussed more fully in Biet et al. [2]. We performed a preliminary user study and found that the squeeze film effect between the glass plate and the finger starts to be felt at a vibration amplitude of around 1.5 microns.

7.3 Psychophysical Evaluation

A friction discrimination experiment is implemented. Subjects are asked to explore the display surface with their bare index finger and describe to the researchers the sensations they feel. The friction discrimination experiment determines the just noticeable difference (JND) of the friction coefficient for humans interacting with the TPaD.

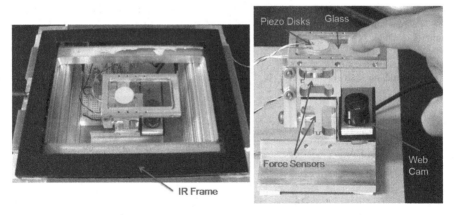

Fig. 7.3 The TPaD and the experimental setup. While a subject is interacting with the TPaD as shown on the *right*, finger position and normal and friction forces are recorded

7.3.1 Friction Force Discrimination

7.3.1.1 Experimental Setup

During the experiments, the subject's finger position was tracked by a commercial infrared (IR) touch screen frame (see Fig. 7.3). A commercial web camera was also used to record the finger tip. In addition to the finger tip position recording, normal and translational forces applied by the finger were measured by two 0.25 and 0.5 lb capacity load cells (type: LSM250, FUTEK Advanced Sensor Technology, Inc.) attached to the TPaD device. The normal and friction forces were collected throughout the trial at a sampling rate of 1000 Hz. The finger position was recorded at 100 Hz.

7.3.1.2 Experimental Method

The friction discrimination experiment determines the just noticeable difference (JND) of the friction coefficient for humans interacting with the TPaD. The experiment requires the subject to compare the slipperiness of two stimuli, displayed in sequential order. A baseline value of the coefficient of friction is always one of the two stimuli presented in the discrimination task. The other stimulus is a test value to compare against the baseline. The subject explores two stimuli and chooses the one with higher coefficient of friction. He/she can also choose the answer "same/can't tell". In all experiments, the subject interacts with the TPaD shown in Fig. 7.3.

The first test value is chosen to be $5\Delta V$ higher than the baseline. The next trial's test value is adjusted according to Kaernbach's adaptive staircase method [6] until a final JND is found (for details on the method, see Chap. 3). If the subject correctly identifies which stimulus has a higher coefficient of friction, the excitation voltage

Fig. 7.4 When the TPaD is OFF, the coefficient of friction is around 0.95. Even $10V$ peak-to-peak excitation (corresponding to a friction coefficient of $\mu = 0.7$) creates a perceivable difference. Subject responds "different" and correctly identifies the stimulus with a higher coefficient of friction

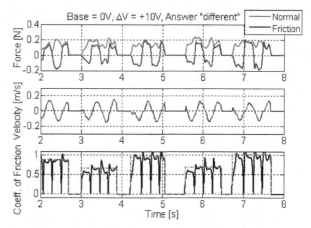

of the test stimuli for the next trial is brought closer to the baseline stimulus by ΔV. If the subject is incorrect, the difference between the baseline and next trial's test stimulus is increased by $3\Delta V$. If the subject answers that the stimuli are the "same/can't tell", the difference between the baseline and next trial's test stimulus is increased by ΔV. These quantities were determined using Kaernbach's formula given in Eq. (3.3) for a target performance of 75 % correct. The JND is reached when the subject has performed eight reversals within a baseline set. A reversal is when the direction of adjusting ΔV changes. The difference threshold for the particular baseline stimulus is evaluated by averaging the test stimuli values between the 4th and 8th reversal.

A set of trials takes place for each baseline excitation voltage. Different baseline values are necessary since JND values for friction coefficients might depend on the magnitude of the friction coefficient. The baseline values for excitation voltages (peak-to-peak) on the TPaD are 0, 10, 20, 30, 40, 50, 60V corresponding to the coefficient of friction of 0.95, 0.7, 0.6, 0.4, 0.3, 0.2, 0.17 respectively. The step size, ΔV, is chosen as $5V$.

During each trial the subject moves his/her finger back and forth on the disk, attempting to maintain a constant normal force and velocity. Throughout each trial the subject is free to toggle between the baseline and the test stimuli as many as he/she wants. The finger is lifted off the plate at each toggle.

7.3.1.3 Results & Discussion

Preliminary data from one subject is presented. A total of 96 data collection trials were performed. An example of data obtained during a trial is shown in Fig. 7.4. The data is from a whole trial with 4 toggles between the baseline stimulus (excitation voltage $0V$, i.e., TPaD is OFF) and the test stimulus ($10V$ peak-to-peak). The coefficient of friction (μ) between the finger and the plate surface was calculated using the formula of Coulomb friction. The dashed line in the 3rd row of Fig. 7.4

Fig. 7.5 30*V* and 45*V* correspond to the coefficients of friction of $\mu = 0.35$ and 0.2, respectively. The subject perceives the difference of 0.15 in the coefficient of friction

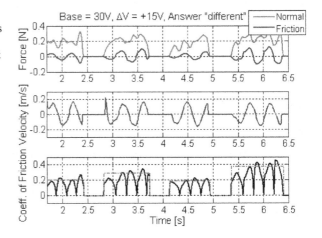

Fig. 7.6 At the baseline value of $\mu = 0.3$ (corresponding to 40*V*), 0.05 difference in the coefficient of friction (5*V*) is not distinguishable. Subject responds "same"

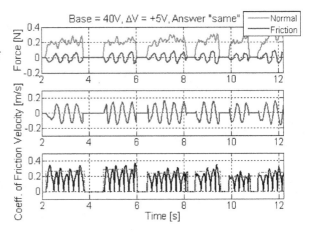

shows the mean coefficient of friction which is averaged for each stimulus after a threshold is applied. As seen in the figure, the coefficient of friction is around 0.95 when TPaD is OFF and 10*V* peak-to-peak results in 0.25 reduction of the friction coefficient value. The subject responded "different" after exploring the two stimuli. Therefore, we can conclude that the 10*V* excitation reduces the friction perceivably.

Figures 7.5 and 7.6 show two cases in which the subject responded "different" and "same", respectively. In Fig. 7.5, the baseline value of the coefficient of friction is 0.35 (excitation voltage is 30*V*) and it reduces to 0.2 when 45*V* is applied. The subject perceives the difference of 0.15 in the coefficient of friction at this baseline value. On the other hand, a 0.05 difference in the coefficient of friction is not recognizable with respect to the baseline value of 0.3 (40*V*) as shown in Fig. 7.6.

The coefficient of friction for each excitation voltage is shown in Fig. 7.7. The data points represent the mean values of the coefficient of friction for each excitation voltage and the vertical bars correspond to the standard deviations in the coefficient

Fig. 7.7 Reduction of friction as a function of excitation voltage. The *data points* represent the mean values of the coefficient of friction derived from each stimulus and the *vertical bars* correspond to the standard deviations over stimuli. The *dashed line* represents curve fitting to the *data points*

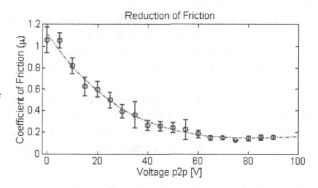

Fig. 7.8 Just noticeable difference in friction represented as a Weber Fraction and plotted versus baseline coefficient of friction (stimulus intensity). The Weber Fraction appears to be relatively independent of stimulus intensity

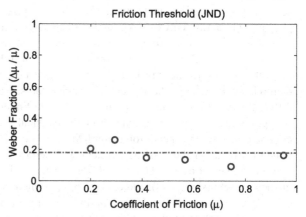

of friction over stimuli. As shown in the figure, reduction of friction starts at $10V$ peak-to-peak excitation voltage and reaches a limit above $70V$.

The data for this one subject may be summarized as a plot of Weber Fraction versus baseline level, as shown in Fig. 7.8. The average JND of friction is 18 %. This is similar to that of other haptic modalities such as compliance (22 % as reported in Tan et al. [11]), force and viscosity (7 % and 34 %, respectively, as reported in Jones and Hunter [5]). This result is encouraging, suggesting that friction modulation may be successfully used to present haptic effects to the bare finger, just as compliance and viscosity modulation have been used to present haptic effects to the whole hand. Of course, tests with additional subjects are necessary before firm conclusions can be drawn.

7.4 Conclusions & Future Work

The recorded data in this experiment (e.g. area of contact, applied tangential and normal forces) can also be used to examine the exploratory behavior of subjects and the finger conditions (humidity, skin type etc.) in order to better understand how humans perceive rendered textures. We are currently carrying out the friction discrimination experiment with additional subjects.

The friction discrimination experiment was based on the force discrimination test explained in Chap. 6. In fact, all the tests, including the ones related to motor control mode, can be applicable to the TPaD to some extent. The reason why we haven't implemented these tests for the TPaD is that the friction modulation capability of the TPaD is not known explicitly yet. Before implementing more sophisticated tests, basic experiments should be conducted. Therefore, in addition to the friction discrimination experiment, we are planning to perform two other psychophysical experiments:

Magnitude Estimation of Friction This experiment will quantify suprathreshold sensitivity to variations in friction. The ultimate goal is to understand how the physical interaction between skin and surface gives rise to the perception of friction. The subject will explore a surface and give a free numerical rating of its friction intensity. A magnitude-estimation function, relating the mean rated perceptual intensity to the physical dimension, will be obtained (for details on the method, see Chap. 3). The subject will repeat the same procedure with applying different normal forces and exploring at different finger velocities. We will relate changes in the percept of friction to both the physical variation and the exploratory parameters including speed, force and RMS deviation from a straight line.

Shape Identification The last experiment will determine the range of shapes that may be conveyed by the TPaD, with respect to spatial extent (size), spatial frequency content (sharpness), and exploratory parameters. This experiment will also quantify the threshold for discriminating a grating from a smooth surface. The subject will explore rendered Gaussian shapes (virtual "bumps" and "holes") and be instructed to locate the highest/lowest point of the highest/deepest perceived bump/hole. Data recorded will include final position of the finger, normal and tangential forces and the subject's shape identification response.

References

1. Biet, M., Giraud, F., Martinot, F., Semail, B.: A piezoelectric tactile display using traveling lamb wave. In: Proc. of Eurohaptics'06 (2006)
2. Biet, M., Giraud, F., Lemaire-Semail, B.: Squeeze film effect for the design of an ultrasonic tactile plate. IEEE Trans. Ultrason. Ferroelectr. Freq. Control 54(12), 2678–2688 (2007)
3. Biet, M., Giraud, F., Lemaire-Semail, B.: Implementation of tactile feedback by modifying the perceived friction. Eur. Phys. J. Appl. Phys. 43, 123–135 (2008).
4. Chubb, E.C., Colgate, J.E., Peshkin, M.A.: Shiverpad: a glass haptic surface that produces shear force on a bare finger. IEEE Trans. Haptics 3, 189–198 (2010).
5. Jones, L.A., Hunter, I.W.: A perceptual analysis of viscosity. Exp. Brain Res. 94, 343–351 (1993)
6. Kaernbach, C.: Adaptive threshold estimation with unforced-choice tasks. Percept. Psychophys. 63(8), 1377–1388 (2001)
7. Marchuk, N.D., Colgate, J.E., Peshkin, M.A.: Friction measurements on a large area tpad. In: Proc. of the Haptics Symposium, pp. 317–320 (2010)

8. Nara, T., Takasaki, M., Maeda, T., Higuchi, T., Ando, S., Tachi, S.: Surface acoustic wave tactile display. IEEE Comput. Graph. Appl. **21**, 56–63 (2001)
9. Robles-De-La-Torre, G., Hayward, V.: Force can overcome object geometry in the perception of shape through active touch. Nature **412**, 445–448 (2001)
10. Samur, E., Colgate, J.E., Peshkin, M.A.: Psychophysical evaluation of a variable friction tactile interface. In: Proc. of Human Vision and Electronic Imaging XIV, vol. 7240 (2009)
11. Tan, H.Z., Durlach, N.I., Beauregard, G.L., Srinivasan, M.: Manual discrimination of compliance using active pinch grasp: the roles of force and work cues. Percept. Psychophys. **57**(4), 495–551 (1995)
12. Watanabe, T., Fukui, S.: A method for controlling tactile sensation of surface roughness using ultrasonic vibration. In: IEEE International Conference on Robotics and Automation, vol. 1, pp. 1134–1139 (1995)
13. Winfield, L., Glassmire, J., Colgate, J.E., Peshkin, M.: T-PaD: Tactile pattern display through variable friction reduction. In: Second Joint Eurohaptics Conference and Symposium on Haptic Interfaces for Virtual Environment and Teleoperator Systems, pp. 421–426 (2007)

Part IV
Synthesis of Evaluation Methods

Chapter 8
Conclusion

Abstract A combined physical and psychophysical experimental methodology is proposed in this book. First, the existing physical performance measures and device characterization techniques were investigated and described in an illustrative way. The physical characterization methods were demonstrated on a two degrees-of-freedom haptic interface. Second, a wide range of human psychophysical experiments were reviewed and the appropriate ones were applied to haptic interactions. The psychophysical experiments were unified as a systematic and complete evaluation method for haptic interfaces. Seven psychophysical tests were derived and implemented for three commercial force-feedback devices. Experimental user studies were carried out and applicability of the tests to a tactile feedback device was investigated. This chapter summarizes the contributions and provides a synthesis of the physical and psychophysical evaluation methods.

8.1 Summary

The main goal of this book was to establish a norm for haptic interface evaluation and identify significant benchmark metrics. As a first step towards this goal, the experimental evaluation was divided into two as physical characterization and psychophysical evaluation. In the first part, the physical performance metrics in the literature were investigated. The variety of the measures, their dependency on boundary conditions and testing methods were discussed in order to show the challenges in device evaluation and comparison. Then, we applied the revised methods to a novel two degrees-of-freedom haptic interface to identify possible improvements in the design of the device.

In the second part, we proposed a set of evaluation testbeds derived from psychophysical experiments that examine haptic interactions. A wide range of different tasks were considered and unified in a standard method. The taxonomy of the haptics mode was used as a starting point for evaluation. Based on this taxonomy, we defined the generic haptic interaction tasks for perception (detection, discrimination, and identification). These testbed modules can also be seen as a complementary work to Bowman and Hodges' [1] testbeds on travel, selection and manipulation. In the testbeds, performance results are measured in terms of information transfer expressed in bits and sensory (absolute and differential) thresholds which can be used

E. Samur, *Performance Metrics for Haptic Interfaces*,
Springer Series on Touch and Haptic Systems,
DOI 10.1007/978-1-4471-4225-6_8, © Springer-Verlag London 2012

as benchmark metrics to evaluate and compare different haptic interfaces. Seven testbeds were designed and a standard VE experimental setup was implemented for three commercially available force-feedback devices. Then we carried out user studies and presented resulting benchmark metrics for these interfaces. The results proved that the testbeds were suitable to compare the three force feedback interfaces in terms of the proposed performance metrics. The differences in the metrics revealed the basic characteristics of the haptic interfaces which are not easily derived from the given specifications. The results were also compared with those from human psychophysical experiments with real objects. These analyses showed significant differences between the experiments performed by the haptic devices and the human psychophysical tests with real objects. This clearly demonstrates the limits of state-of-the-art haptic interfaces and the need for high fidelity haptic interfaces.

In this book, we also investigated the validity of the experimental approach for tactile feedback devices. Friction discrimination experiments were performed with a tactile pattern display. Preliminary results proved that the device is capable of modulating friction to present haptic effects to the bare finger successfully. Also, it was shown that the testbeds related to the both motor control and perception mode of haptics can well be applied to the tactile devices. In addition to the testbeds, two other psychophysical tests were also proposed for tactile device evaluation.

8.2 Synthesis of Physical and Psychophysical Methods

Considering the proposed seven psychophysical testbeds and defined numerous physical experiments, to run all the tests is exhaustive. However, a synthesis of the two proposed evaluation approach may reduce the number of tests that need to be run to characterize a device. Table 8.1 shows how the physical and psychophysical metrics are related to and complement each other. As the psychophysical testbeds are intuitively connected to the aim of a task, users may decide which testbed and metrics they should consider for device selection depending on the target application. They can also take into account its corresponding technical specification referring to this table. For instance, if the aim of a target application is to manipulate objects in free space, results of first two experiments should be considered since they involve the tasks (travel, selection and manipulation) related to the motor control mode of haptics. These results are mainly influenced by the kinematics and also the minimum output impedance of an interface. Therefore, the peg-in-hole test (i.e., the second test), which comprises also the tapping test, can be used to determine the level of transparency instead of the physical impedance experiments. This human-in-the-loop experiment presents better measurement conditions which highly affect the stability of a system.

The force perception experiments are purely related to the actuation system properties. Therefore, if a calibration curve is obtained through static physical measurements, the force detection and discrimination tests may be redundant. However, the force discrimination testbed may still provide additional information since the measurements are performed under dynamic conditions and at different locations. This

Table 8.1 Synthesis of physical and psychophysical metrics

Psychophysical Testbeds			Physical Evaluation			
Aim	#	Metric	Kinematics	Actuation	Sensing	Impedance
Motor Control	1	IP, a	Workspace		Min	
	2		DOF			Max
Force Perception	3	Abs. threshold		Min force		
	4	Force JND		Resolution	Max force	
Texture Perception	5	Position JND		Bandwidth	Resolution	
Geometry Perception	6	IT	Structure			Max
	7		DOF			

enables us to test the uniformity of force distribution over the workspace. On the other hand, the performance results of a texture experiment provide more valuable information combining both actuation and sensing properties such as force bandwidth and position resolution. For example, one may rely on position resolution specifications to render fine spatial gratings according to the reconstruction limit equation defined by Campion and Hayward [2]. Using this equation we might conclude that the device has an excellent grating rendering property. However, it would not be correct in real implementation since the frequency response of the system highly influence simulation of gratings. A recent study by Cholewiak et al. [3] shows this relation. However, it is still difficult to relate the bandwidth and resonant frequencies to the rendered gratings. Therefore, a texture discrimination experiment is useful to obtain some information about what can actually be rendered by a device.

The structure and maximum impedance characteristics of a device influence the perception of a geometry. However, technical specifications related to these characteristics do not provide an intuitive understanding on whether a device supports the geometric identification or not. Apart from the workspace, structure type and DOF, no geometrical specification is provided among these specifications. On the other hand, the performance metric IT is a quantitative measure providing the degree of ability to identify shapes and sizes using a device. It is also possible to combine shape and size experiments into a single identification experiment to reduce the number of tests. For instance, the curvedness of the quadric shapes could be altered.

In addition to above three crucial psychophysical tests (the peg-in-hole, the texture discrimination and the shape identification), three physical evaluation methods should be performed for full device characterization: the static response measurements for actuation and sensing and the impedance range (i.e., Z-width) measurements. First two will give the input-output curves and the latter will define the device's ability to render a wide range of haptic stimuli. Overall, these six physical and psychophysical tests complement each other and constitute a holistic device characterization suite.

8.3 Contributions

In order to identify guidelines for device characterization, physical performance measurements for haptic interfaces were described in an illustrative way. Although various performance metrics were defined in the literature, it was almost not possible to find detailed information on specifications, testing conditions and methods. Tutorial-like guidelines for physical device evaluation described in this book provides sufficient information in detail to perform identical tests with other devices.

In terms of evaluation, a generic and reusable validation method which comprehensively addresses all attributes of haptic interaction was developed. The main contribution of this book, first, is providing new quantitative performance metrics in addition to the available specifications of the haptic interfaces. The main benefit of these quantitative metrics is that it is easier to link them to the rendering quality perceived by the user. They also allow us to show how different performances of the devices are in a given task. Since the human user is in the loop not only physically but also cognitively during the evaluation process, these tests help us to determine actual or probable usefulness of a device with quantifiable characteristics. In addition, they provide a better understanding of the sensory resolution associated with a device. Considering human perceptual thresholds as a norm for meaningful device comparison and assessment, this study shows detailed device-specific limits. Since not all the manufacturers provide detailed information on specifications, these limits provide useful information about the devices that cannot be otherwise predictable straightforwardly.

In the experimental setup, the same virtual environment system should be used in order to make a comparison between different interfaces. Therefore, a standard VE experimental setup was developed. The only variation is the between-device differences and all the other factors were controlled. Therefore the experimental results show a fair comparison of three devices for basic haptic interaction tasks. In the literature, one can find similar studies for one of the tests or for one device, but since there is no across-study consensus, it is not possible (or would be misleading) to compare the results. Therefore, we believe that our unified approach covering all aspects of haptic interaction for three devices is a considerable contribution to the state of the art. In addition, we paid careful attention to the design of experiments to identify the exhaustive variables with a minimum number of experiments but still in a statistically appropriate way.

8.4 Outlook

The methodology proposed here is a first step towards a standardized evaluation method of tactile and force feedback devices which is also parallel to ISO's work on international standards for haptic and tactile interactions [5]. For this reason, in the future we will develop an open source library available to the haptic community in order to apply designed testbeds to other types of haptic interfaces.

Evaluations conducted in a generic context can be applied to many types of haptic interfaces. However, application-specific evaluation methods should be conducted in order to evaluate how well an application's haptic interface supports correspondent user tasks. For instance, the Xitact IHP, which is designed for the laparoscopy simulation, should be evaluated in terms of complete system and operation. In other words, clinical tests should validate the device's proper use in the context of surgical simulation.

Although the physical measurements were described in detail, there is still the challenge of making the measurements in practice. Therefore, an experimental workbench with capabilities of measuring all the metrics is being developed to obtain standardized evaluation.

The results of the psychophysical tests are highly related to the device characteristics. However, in-depth analysis are still required to understand how each physical parameter affects the perception as it has been performed for vibration perception in [4]. To realize this, psychophysics alone will likely not be sufficient. Instead, combined psychophysical, physiological and also neurophysiological analysis are required. This research would result in guidelines that are psychophysically/physiologically driven.

References

1. Bowman, D.A., Hodges, L.F.: Formalizing the design, evaluation, and application of interaction techniques for immersive virtual environments. J. Vis. Lang. Comput. **10**(1), 37–53 (1999)
2. Campion, G., Hayward, V.: Fundamental limits in the rendering of virtual haptic textures. In: World Haptics Conference, pp. 263–270 (2005)
3. Cholewiak, S.A., Kim, K., Tan, H.Z., Adelstein, B.D.: A frequency-domain analysis of haptic gratings. IEEE Trans. Haptics **3**, 3–14 (2010)
4. Salisbury, C.M., Gillespie, R.B., Tan, H.Z., Barbagli, F., Salisbury, J.K.: What you can't feel won't hurt you: evaluating haptic hardware using a haptic contrast sensitivity function. IEEE Trans. Haptics **4**(2), 134–146 (2011)
5. van Erp, J.B.F., Kern, T.A.: ISO's work on guidance for haptic and tactile interactions. In: Haptics: Perception, Devices and Scenarios, vol. 5024, pp. 936–940 (2008)

Glossary

Absolute threshold Minimum amount of stimulation required for a human to detect a stimulus.

Active DOF Number of independent force feedback directions.

Detection experiment Measurement of absolute sensory thresholds.

Dexterity Ability to move and apply forces and torques in arbitrary directions.

Dexterous workspace Points in a workspace that can be reached by arbitrary orientations.

Differential threshold Smallest difference in a stimulus detectable by a human.

Discrimination experiment Measurement of differential sensory thresholds.

Dynamic range Ratio of the maximum value of a parameter to the minimum.

Elastostatics Study of elasticity under static equilibrium conditions.

Force resolution Smallest incremental force that can be generated in addition to the minimum force.

Haptic interface An actuated, computer controlled and instrumented device that generates the sense of virtual touch in the form of force feedback (for receptors in the muscles and joints) and/or tactile feedback (for sensors located in the skin).

Haptic technology Technology that deals with the synthesis of touch and force to enable users to interact with virtual or real environments through haptic interfaces.

Identification experiment Measurement of human ability to categorize stimuli.

Impedance Relation between velocity and force.

Impedance range Dynamic range of impedances that can be passively rendered with a haptic interface, i.e., Z-width.

Index of difficulty An index expressed in bits that shows the difficulty level in the Fitts' law.

Index of performance An index expressed in bit/s that represents the performance rate in a task.

Manipulability Ease of arbitrarily changing position and orientation for a given posture.

Manipulation Task of setting the position and orientation of a virtual object in a virtual environment.

E. Samur, *Performance Metrics for Haptic Interfaces*,
Springer Series on Touch and Haptic Systems,
DOI 10.1007/978-1-4471-4225-6, © Springer-Verlag London 2012

Maximum continuous force Maximum amount of force that can be generated by a haptic interface which is not limited in time.

Maximum impedance Highest possible gain that can be rendered by a haptic interface without any human induced control instabilities by voluntary motion.

Mechanical impedance Resistance of a system to a motion.

Minimum force Minimum amount of force that can be generated by a haptic interface.

Minimum impedance Relation between the least resisting force for a given velocity input when a free space is simulated in a virtual environment.

Operating bandwidth Frequency which determines the useful range within which a system can operate.

Output impedance Dynamic relation between a given velocity input and resulting force provided by a haptic interface.

Passive DOF Number of independent motion directions driven by the user.

Position resolution Smallest change of position which can be detected by sensors.

Reachable workspace Set points in a workspace that can be reached by an end effector.

Selection Act of choosing a virtual object in a virtual environment.

Sensitivity Slope of a calibration curve, i.e., amount of change in output for a unit change in input.

Travel Movement of one's viewpoint in a virtual environment.

Transparency Quality of reflecting desired impedance through a haptic interface.

Index